THE ILLUSTRATED FLORA OF ILLINOIS

The Illustrated Flora of Illinois

ROBERT H. MOHLENBROCK, General Editor

THE ILLUSTRATED FLORA OF ILLINOIS

GRASSES
bromus to paspalum

Robert H. Mohlenbrock

SOUTHERN ILLINOIS UNIVERSITY PRESS
Carbondale and Edwardsville

FEFFER & SIMONS, INC.
London and Amsterdam

Copyright © 1972 by Southern Illinois University Press
All rights reserved
Printed in the United States of America
Designed by Andor Braun
Illustrations by Miriam W. Meyer
International Standard Book Number 0–8093–0520–8
Library of Congress Catalog Card Number 71–156793

This book is dedicated to
Professor John W. Voigt,
colleague and ardent student of grasses,
who guided me through my first years of graduate work,
and who has been a cherished friend since

CONTENTS

ILLUSTRATIONS

FOREWORD

The grasses of Illinois will appear in two volumes of The Illustrated Flora of Illinois series. It follows publication of the ferns of Illinois and two volumes on monocotyledonous plants of Illinois. The series grew out of an idea to present all information known about every kind of plant which occurs in Illinois. The Illustrated Flora of Illinois is a multivolumed flora of the state of Illinois, to cover every group of plants, from algae and fungi through flowering plants. In addition to keys and descriptions of every plant known to occur in Illinois, there would be provided illustrations showing the diagnostic characters of each species.

An advisory board was set up in 1964 to screen, criticize, and make suggestions for each volume of The Illustrated Flora of Illinois during its preparation. The board is composed of taxonomists eminent in their area of specialty—Dr. Gerald W. Prescott, University of Montana (algae); Dr. Constantine J. Alexopoulos, University of Texas (fungi); Dr. Aaron J. Sharp, University of Tennessee (bryophytes); Dr. Rolla M. Tryon, Jr., The Gray Herbarium (ferns); Dr. Robert F. Thorne, Rancho Santa Ana Botanical Garden and Mr. Floyd Swink, the Morton Arboretum (flowering plants).

This author is editor of the series and will prepare many of the volumes. Specialists in various groups are preparing the sections of their special interest.

There is no definite sequence for publication of The Illustrated Flora of Illinois. Rather, volumes will appear as they are completed.

Robert H. Mohlenbrock

Southern Illinois University
December 1971

The Illustrated Flora of Illinois

GRASSES
bromus to paspalum

County Map of Illinois

Introduction

The nomenclature for species followed in this volume is based largely on that of Hitchcock (1950) in the *Manual of the Grasses of the United States,* except where recent monographs and revisions are available. The division of the grass family into subfamilies and tribes essentially follows Gould (1968) and is a major departure from the sequence usually found in most floristic works in North America.

Synonyms, with complete author citation, which have applied to species in the northeastern United States, are given under each species. A description, based primarily on Illinois material, is provided for each species. The description, while not necessarily intended to be complete, covers the more important features of the species.

The common name (or names) is the one used locally in Illinois. The habitat designation is not always the habitat throughout the range of the species, but only for it in Illinois. The overall range for each species is given from the northeastern to the northwestern extremities, south to the southwestern limit, then eastward to the southeastern limit. The range has been compiled from various sources, including examination of herbarium material. A general statement is given concerning the range of each species in Illinois. Dot maps showing county distribution of each grass in Illinois are provided. Each dot represents a voucher specimen deposited in some herbarium. There has been no attempt to locate each dot with reference to the actual locality within each county.

The distribution has been compiled from field study as well as herbarium study. Herbaria from which specimens have been studied are the Field Museum of Natural History, Eastern Illinois University, the Gray Herbarium of Harvard University, Illinois Natural History Survey, Illinois State Museum, Missouri Botanical Garden, New York Botanical Garden, Southern Illinois University, the United States National Herbarium, the University of Illinois, and Western Illinois University. In addition, a few private collections have been examined.

Each species is illustrated, showing the habit as well as some of the distinguishing features in detail. Most of the illustrations have been prepared by Miriam Wysong Meyer. Dr. Kenneth

Lewis Weik illustrated 121, 125, 127, 129, 130, 131, 132, 133, 147, 149, and 171. Fredda Burton prepared figures 4, 9, 10, 21, 22, 26, 79, 80, 91, and 102.

Several persons have given invaluable assistance in this study. Mr. Floyd Swink of the Morton Arboretum has read and commented on the entire manuscript. For courtesies extended in their respective herbaria, the author is indebted to Dr. Robert A. Evers, Illinois Natural History Survey; the late Dr. G. Neville Jones, University of Illinois; Dr. Glen S. Winterringer, Illinois State Museum; Dr. Arthur Cronquist, New York Botanical Garden; Dr. Jason Swallen, the United States National Herbarium; Dr. Loren I. Nevling, the Gray Herbarium; Dr. George B. van Schaack, formerly of the Missouri Botanical Garden and now of the Morton Arboretum; and Dr. Walter Lewis, the Missouri Botanical Garden.

Southern Illinois University provided time and space for the preparation of this work. The Graduate School and the Mississippi Valley Investigations and its director, the late Dr. Charles Colby, all of the Southern Illinois University, furnished funds for the field work and the salaries for the illustrators.

HISTORY OF GRASS COLLECTING IN ILLINOIS

There always have been many kinds of grasses known from Illinois. Mead, who published the first extensive list of Illinois plants in 1846, recorded 83 species of grasses. Lapham, who wrote specifically about Illinois grasses in 1857, listed some 135 species. Patterson, in his catalog of Illinois plants nineteen years later (1876), reduced this number to 123 species.

The most important work on Illinois grasses has been by Mosher, in 1918, when she prepared the *Grasses of Illinois,* a treatise providing descriptions and cited specimens of all grasses known to occur in the state. Two hundred and four species are recorded in her work.

Jones reported 212 species of grasses in 1945 and 220 in 1950. Jones, Fuller, Winterringer, Ahles, and Flynn added 26 species in 1955, bringing the total to 246. In 1960, Winterringer and Evers included 8 additional species of grasses from Illinois. Glassman has studied the grasses of the Chicago region thoroughly for the past several years, and his treatment of these (1964) is excellent.

During the research for this book, several species of grasses

previously unreported from Illinois were found in various herbaria, for the most part bearing misidentifications. A number of
additional species was found during intensive field study, particularly in the southern one-third of the state. Differences in the taxonomic treatment have accounted for the addition or subtraction
of some species within the state. In these volumes on grasses 286
species are recognized from Illinois, along with 49 lesser taxa.

MORPHOLOGY OF GRASSES

Grasses belong to the family Poaceae (also called Gramineae).
Until recently, most botanists grouped grasses and sedges (Cyperaceae) in the order Graminales (or Poales). Anatomical, morphological, and other more recent evidence show that, in addition
to grasses and sedges, other families such as the Xyridaceae, Commelinaceae, Pontederiaceae, and Juncaceae share some of the
same characters and should be grouped together. This view is
followed here so that these six families are considered to comprise the Commelinales. The Xyridaceae, Commelinaceae, Pontederiaceae, and Juncaceae are treated in *Flowering Plants:
Flowering Rush to Rushes* (1970); the Cyperaceae will be forth-
coming in two subsequent volumes.

The nature of grass structures generally is so different from
that of other flowering plants that a special terminology is applied
to grasses. A thorough understanding of these terms will enable
one to identitify more readily an unknown specimen.

Grasses are annuals, biennials, or perennials. Annuals have
tufts of fibrous roots and live for a single growing season. Perennials may be tufted (*Fig. 1*), or they may have rhizomes (horizontal, root-producing stems below ground [*Fig. 2*]), or they may

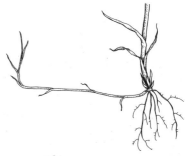

1. Tufted perennial.   *2.* Rhizome.

have stolons (horizontal, root-producing stems above ground [*Fig. 3*]), or a short, thick, subterranean crown (*Fig. 4*).

The stem which bears the leaves and inflorescence is called the culm. While the culm may be hollow or solid, the nodes (where the leaves arise) are nearly always solid. The culms may be simple or branched. Often they are jointed (geniculate) near the base (*Fig. 5*). Culms may be erect, divergent (spreading), or prostrate and matted.

Grass leaves are borne at the nodes in two planes along the culm. This condition is referred to as 2-ranked (*Fig. 6*). Sometimes, because of a twisting of the culm, the 2-ranked condition is not apparent. The leaf is composed of a blade and a sheath. The sheath wraps around and encloses a portion of the culm. If the margins of the sheath are united, forming a cylinder, the sheath is closed (*Fig. 7*); if the margins are not united, the sheath is open (*Fig. 8*). The blade is the free portion of the leaf. It is parallel-veined and generally elongated, although some grasses with rather short, broad blades occur. The blades normally are flat, but they may be folded (plicate [*Fig. 9*]) or inrolled into a slender tube (involute [*Fig. 10*]). Along the inner face of the leaf, where the blade adjoins the sheath, there is often a ciliate, membranous, or cartilaginous structure of varying size and shape known as a ligule (*Fig. 11*). In some grasses, some of the leaves are not blade-bearing, therefore consisting merely of sheaths.

The inflorescence is the aggregation of a group of spikelets (the basic unit of the grass inflorescence). An elongated, simple axis with pedicellate spikelets borne along it is called a raceme (*Fig. 12*); if the spikelets are sessile along the simple axis, the inflorescence is a spike (*Fig. 13*). Short-pedicellate spikelets crowded on an elongated, simple axis make up the spike-like raceme (*Fig. 14*). If the inflorescence is branched, and the spike-

3. Stolon.   4. Short, thick,   5. Geniculate
                  subterranean crown.   base of stem.

6. Two-ranked leaves.

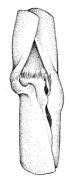

7. Closed sheath.

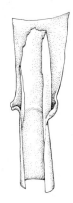

8. Open sheath.

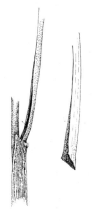

9. Plicate leaf.

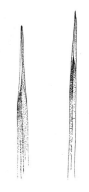

10. Involute leaf.

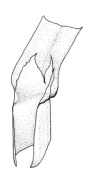

11. Ligule.

12. Raceme.

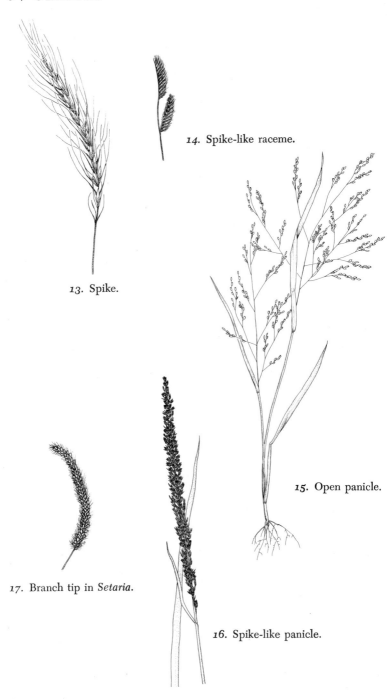

14. Spike-like raceme.

13. Spike.

15. Open panicle.

17. Branch tip in *Setaria*.

16. Spike-like panicle.

lets are pedicellate, the term used is panicle (*Fig. 15*). The panicle may be very wide-spreading and open (diffuse [*Fig. 15*]), or it may be contracted so much as to resemble a spike (*Fig. 16*). This latter situation gives rise to the term spike-like panicle.

The tip of each branch of the panicle normally bears a spikelet, although in *Setaria* (*Fig. 17*) and *Cenchrus* (*Fig. 18*), some of the branch tips are sterile and modified into bristles.

The spikelet is composed of an axis, called the rachilla, along which are borne bracts in two ranks (*Fig. 19*). The lowest two bracts bear no flowers in their axils. These "empty" bracts are the glumes. They frequently are unequal in size although rarely unlike in texture. Both glumes are essentially lacking in *Leersia* and *Zizania*, while the first (lower) glume is usually absent in *Paspalum, Digitaria, Eriochloa,* and *Lolium. Elymus hystrix* usually has its glumes reduced to awns. A sharp ridge down the back of a compressed glume is called the keel (*Fig. 20*). Sometimes the entire spikelet falls at maturity, while in other species the glumes remain behind. In the first case, the spikelet is said to disarticulate below the glumes, while in the latter case, it is said to disarticulate above the glumes.

Above the glumes are one or more bracts which usually bear a flower within. These fertile bracts are the lemmas. Facing each lemma is a usually somewhat smaller palea (*Fig. 21*). Between the lemma and the palea is the flower (*Fig. 22*). In *Chasmanthium* and *Panicum*, the lowest lemma does not produce a flower, while in *Melica* and *Chloris*, the uppermost lemma is sterile. In *Phalaris*, the two lowest lemmas are reduced to scales. Lemmas

18. Branch tip in *Cenchrus.*

19. Spikelet with bracts in two ranks.   20. Keel on glume.

generally are of the same texture as the glumes, although the fertile lemma in *Panicum* is indurated. The callus of a lemma may refer to a swollen, hardened area at its base (as in *Stipa* and *Aristida* [Fig. 22]) or a tuft of hairs (as in *Calamagrostis* [Fig. 23]). Lemmas also may be keeled. Spikelets with a single fertile lemma are said to be 1-flowered (*Fig. 24*), while those with two or more fertile lemmas are several-flowered (*Fig. 25*).

The palea is smaller than the lemma and usually of more delicate texture. In *Panicum hians,* the palea becomes indurated at maturity. The palea is often absent in *Agrostis.* Most paleas have two keels down the back.

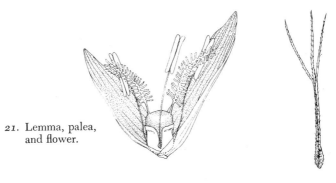

21. Lemma, palea, and flower.

22. Callus at base of lemma.

24. One-flowered spikelet.

23. Tuft of hairs at base of lemma.

The grass flower is much reduced from the flower of Liliaceae and other more showy flowering plants. It consists of three stamens (occasionally 1–6) and one pistil. Each stamen bears a 2-celled anther. Each pistil is 1-celled, with but one ovule, but there usually are 2–3 styles. At the base of the flower usually are found 2–3 small scales thought to represent the perianth. These scales are the lodicules (*Fig. 26*).

26. Lodicules.

25. Several-flowered spikelet.

Most grasses have a fruit known as a caryopsis, or grain. The seedcoat of the single seed is united directly to the matured ovary wall (pericarp). (The pericarp is free from the seed in *Eleusine, Crypsis,* and *Sporobolus*). At maturity, the grain drops free from the lemma and palea, or it may fall while enclosed by the lemma and palea.

The lemma, palea, and enclosed flower comprise the floret.

DISTRIBUTION OF ILLINOIS GRASSES

Grasses occur in every possible habitat in Illinois—from standing water to the driest bluff-tops, from prairies to woodlands, from waste places and fields to the deepest canyons. The following discussion of habitats for Illinois grasses is divided into three major sections: moist natural areas, dry natural areas, and waste areas.

Moist Habitats

STANDING WATER   There are few grasses, indeed, which can tolerate partial submergence in water. Those which do occupy this kind of habitat are not common and are very locally distributed. Probably the most widespread aquatic grasses in Illinois are *Glyceria septentrionalis* and *Alopecurus aequalis*. *Zizania aquatica* is found in the northern two-thirds of Illinois, while *Deschampsia flexuosa* is known only from extreme northeastern counties. Predominantly southern species of aquatic grasses are *Paspalum fluitans*, restricted to the southern two-thirds of the state, and *Glyceria arkansana* and *Puccinellia pallida*, which are confined to a single station in the southern tip of Illinois.

MOIST SOIL   In this paragraph will be considered grasses which grow in moist soil, but not generally in woodlands, prairies, or on sandy shores. These are the grasses of low meadows and thickets. The most widespread of these species is *Glyceria striata*, although several others are found locally throughout Illinois. Numbered among these are *Alopecurus carolinianus*, *Muhlenbergia glomerata*, *Leersia lenticularis*, *L. oryzoides*, *Echinochloa walteri*, *E. pungens*, and *Panicum clandestinum*. These species usually occur in considerable abundance where they are found. A few of the moist-meadow species are more common in northern Illinois. These are *Phragmites australis*, *Poa palustris*, *Calamagrostis canadensis*, and *Agrostis alba* var. *palustris*. Other species, such as *Chasmanthium latifolium*, *Paspalum pubiflorum*, *P. laeve*, *Panicum rigidulum*, and *P. anceps* are principally southern. *Arundinaria gigantea* often forms dense thickets (canebrakes) in lowlands in the southern one-fourth of the state.

MOIST SAND   On the sandy shores adjacent to the major waterways of Illinois are a few characteristic grasses. Although several species occasionally occupy this habitat, the most typical are *Eragrostis frankii*, *Paspalum ciliatifolium*, *Cenchrus longispinus*, and *Leptochloa filiformis*. These are grasses which, for the most part, can tolerate the wave action of the larger rivers.

WET PRAIRIES   Wet prairies may be regarded as low, moist, treeless areas with predominantly prairie vegetation. *Spartina pectinata* indicates this type of habitat, although *Sphenopholis obtusata* var. *major* is usually present as well. *Panicum lanuginosum* var. *implicatum* may be found here with some regularity, particularly in the northern counties.

MOIST WOODLANDS   This habitat may occur in a low, rather level terrain, or it may be in the depths of picturesque canyons and ravines. In some cases, scattered boulders may be strewn across the forest floor. A few species seemingly need the protective presence of these boulders for their survival. Species such as *Melica mutica*, *M. nitens*, *Muhlenbergia tenuiflora*, and *M. sylvatica* apparently survive better in the moist, rocky woods. Other species are less dependent on the rocks and grow well in essentially rockless woods. Over a dozen species occur rather commonly in moist woodlands throughout the state. These are *Bromus ciliatus*, *B. pubescens*, *Festuca obtusa*, *Poa sylvestris*, *Elymus hystrix*, *Sphenopholis obtusata* var. *major*, *Cinna arundinacea*, *Muhlenbergia frondosa*, *M. mexicana*, *Agrostis hyemalis*, *Brachyelytrum erectum*, and *Leersia virginica*. These species frequently occur singly and rarely form extensive patches. One species, characteristic of many moist woodlands in northern Illinois, is *Bromus purgans*. Species confined to moist woods in the southern two-thirds of Illinois are *Muhlenbergia brachyphylla*, *M. glabriflora*, *Panicum microcarpon*, *P. polyanthes*, and *P. boscii*.

Dry Habitats (excluding fields)

ROCK LEDGES   Exposed rock ledges, frequently becoming intensively xeric during midsummer, nonetheless may be suitable for the growth of a limited number of species, including some grasses. Characteristic of these xeric ledges are *Vulpia octoflora*, *Agrostis elliottiana*, *Danthonia spicata*, *Panicum gattingeri*, *Sporobolus vaginiflorus*, and *S. neglectus*. In the southern tip of the state *Andropogon virginicus* occurs along these ledges.

DRY SAND   Two distinct areas in which the plants grow in dry sand are found in the northern half of the state. Most extreme is the sand of the dunes along Lake Michigan. The characteristic grasses of this rugged habitat are *Ammophila breviligulata* and *Calamovilfa longifolia*. The other sand habitat is the sandy prairie, such as those studied extensively by Gleason in 1910 in the Hanover, Dixon, and Havana areas. Grasses are a vital component of these sandy areas, serving as sand binders in most instances. Characteristic sand-prairie taxa are *Eragrostis trichodes*, *Stipa spartea*, *Leptoloma cognatum*, *Panicum villosissimum* var. *pseudopubescens*, *Aristida tuberculosa*, *Sporobolus cryptandrus*, *Triplasis purpurea*, *Koeleria macrantha*, *Schizachyrium scoparium*, and *Andropogon gerardii*.

DRY PRAIRIES   The prairies considered in this paragraph are those found neither in low, moist situations nor in sandy areas.

They are of two basic types in Illinois, being situated atop predominantly limestone bluffs or on glacial till (hill prairies), or on generally flat terrain. In either case, the same species usually prevail. Common taxa throughout the state are *Schizachyrium scoparium, Andropogon gerardii, Sorghastrum nutans, Koeleria macrantha, Sporobolus heterolepis,* and *Panicum oligosanthes* var. *scribnerianum.* Of more limited distribution are *Stipa spartea, Bouteloua curtipendula, Panicum perlongum,* and *Andropogon virginicus.*

DRY WOODLANDS   As with species of the moist woodlands, there are some species which seemingly thrive better when boulders are present in the dry woodlands. Those dry, rocky woodland species most characteristic are *Eragrostis capillaris, Muhlenbergia sobolifera,* and *Panicum latifolium.* In the southern one-third of the state, *Panicum dichotomum* var. *barbulatum* becomes an important species of this habitat. In the dry, essentially rockless woodlands, several taxa regularly may be found throughout the state. Included among these are, *Elymus canadensis, E. villosus, E. virginicus, Danthonia spicata, Agrostis perennans, Panicum depauperatum, P. lanuginosum* var. *lindheimeri,* and *P. villosissimum.*

Fields and Waste Ground

Grasses, more than any other plants, seem to have the ability to come into and establish themselves in fields, waste ground, and other open areas. Most grasses which occupy this habitat would be considered weedy. A surprising number of these is native. In the following lists, only those grasses which are common throughout most of the state are considered.

Native Species

Agrostis alba	*Hordeum pusillum*
Agrostis hyemalis	*Muhlenbergia schreberi*
Aristida oligantha	*Panicum capillare*
Aristida ramosissima	*Panicum dichotomiflorum*
Elymus canadensis	*Poa chapmaniana*
Elymus virginicus	*Setaria lutescens*
Eragrostis pectinacea	*Tridens flavus*

Adventive Species

Agropyron repens	*Bromus inermis*
Bromus commutatus	*Bromus racemosus*

Adventive species (*continued*)

Bromus secalinus
Bromus tectorum
Dactylis glomerata
Digitaria ischaemum
Digitaria sanguinalis
Echinochloa pungens
Eleusine indica
Eragrostis cilianensis
Eragrostis poaeoides
Festuca pratensis

Hordeum jubatum
Lolium multiflorum
Lolium perenne
Phleum pratense
Poa annua
Poa compressa
Poa pratensis
Setaria faberi
Setaria viridis
Sorghum halepense

THE RARER GRASSES OF ILLINOIS

In order to qualify for inclusion in this section, a species must not be known from more than three counties in Illinois. This arbitrary cutoff is a little misleading, since a highly specialized habitat restricted to one or two counties may have an abundance of one species which is known from no other area in the state; thus, *Panicum scoparium* is included in the rare category since it is known only from two stations in Pope County, although at one of these stations, many specimens occur. At the other extreme, the record of *Oryzopsis asperifolia* from Illinois rests on a single specimen collected in 1877; this species is almost certainly extinct in the state.

Ninety-five of the Illinois grasses (34%) are known from three counties or less. Of these, 37 species are adventive and are in Illinois through the courtesy of a railroad, a bird, the wind, the highway department, or some other force.

A number of the adventives are extremely surprising in Illinois. One of the most interesting collections of an adventive grass was that of *Trichachne insularis* by John Voigt on October 13, 1954. This species, a native to the West Indies, Central and South America, and from Florida to Arizona, was found by Voigt along a roadside one-half mile south of Cambria in Williamson County. The nearness of the station to Crab Orchard Lake suggests that a water bird may have been responsible for the occurrence of this grass in Illinois.

Perhaps the most remarkable adventive grass in Illinois is *Eriochloa villosa*. This species of eastern Asia, known previously in the United States from Oregon and Colorado, was found in corn and soybean fields near Odell in Livingston County by J. V. Myers and R. A. Evers on August 25, 1950. This species still exists at this station. It also was discovered near Barrington, Cook County, in 1969.

Most of the rare, adventive grasses, having been collected only once or twice, probably no longer exist in Illinois, at least not from their original collection sites. Many of the species have not been seen in Illinois in decades. For example, the crowfoot grass (*Dactyloctenium aegyptium*) has not been found since its original Illinois collection by H. Eggert along a railroad in St. Clair County in 1876; *Poa nemoralis* has not been collected in Illinois since A. B. Seymour found it in Champaign County in 1880; *Bromus brizaeformis* has not been seen in Illinois since its collection by R. Ridgway from Richland County in 1902.

On the other hand, a few of the rare adventives have maintained their existence for years at their original station. One example is that of *Eriochloa contracta*. This western prairie species, collected by John Voigt from a levee bank in Union County in 1954, still exists on the same bank.

Many of the fifty-eight rare native grasses represent species which are at one edge of the range of their distribution. A few others are grasses which simply are rare throughout their entire range.

Some of these species have not been collected for many years and are almost certainly extinct in Illinois. Included here is the most controversial grass in the Illinois flora, *Erianthus brevibarbis*. André Michaux collected the type of this species in 1795. According to Hitchcock (1950), the information on the type specimen reads, "dry hills 5 days distant from the Wabash River toward the mouth of the Missouri." Fernald (1945b) figures that this locality would fall in southern Illinois. This grass has not been found since in Illinois, and is known only from a few stations in Arkansas.

High on the list of native species probably extinct in Illinois is *Schedonnardus paniculatus*, the Tumblegrass. Mead's collections from Hancock County in 1845 are the last for this grass in Illinois. Lost nearly as long from the Illinois flora are *Oryzopsis asperifolia*, collected in 1877 by Shipman from Cook County; *Poa wolfii*, collected at about the same time by Wolf from Fulton County, Brendel from Peoria County, and by Patterson from Henderson County; and *Milium effusum*, during the same era, from Kane and Tazewell counties by Vasey and Brendel, respectively. Clinton's collection in 1892 of *Muhlenbergia* × *curtisetosa* from Champaign County, Eggert's collection in 1893 of *Paspalum dissectum* from Perry County, Pepoon's collection in 1908 of *Bouteloua gracilis* from JoDaviess County, and Hill's collection in 1912 of *Muhlenbergia cuspidata* from Will County, are the last for these species in Illinois.

Other native taxa known from but a single county, along with their original collector, are:

Puccinellia pallida (Torr.) Clausen. LaRue Swamp, Union County, first collected by Julius Swayne in 1951. This species still exists in the LaRue Swamp.

Poa autumnalis Muhl. ex Ell. Jackson Hollow, Pope County, collected by the author in 1963. This species still occurs at this station.

Poa angustifolia L. Pine Hills, Union County, collected by Sharon Poellot in 1967. Several specimens occur at this station.

Poa paludigena Fern. and Wieg. Elgin Swamp, collected by George Vasey in the last half of the nineteenth century.

Glyceria arkansana Fern. LaRue Swamp, Union County, first collected by Bill Bauer in 1940. This species may still be found at this station.

Schizachne purpurascens (Torr.) Swallen. Apple River Canyon, JoDaviess County, collected by F. J. Herman in 1937.

Digitaria villosa (Walt.) Pers. Giant City State Park, Jackson County, collected by the author in 1964. This species still occurs at its original station.

Panicum stipitatum Nash. Swampy ground near West Vienna, Johnson County, collected by the author in 1964.

Panicum longifolium Scribn. Fults hill prairie, Monroe County, collected by James Ozment, Wendell Crews, and the author in 1962.

Panicum hians Ell. Near Gale, Alexander County, collected by the author in 1965. This species still occurs at the original collection site.

Panicum linearifolium Scribn. var. *werneri* (Scribn.) Fern. Starved Rock State Park, LaSalle County, collected by G. N. Jones in 1943.

Panicum nitidum Lam. Devil's Backbone near Grand Tower, Jackson County, collected by James Ozment in 1963.

Panicum subvillosum Ashe. North of Barrington, Lake County, collected by James Ozment in 1964.

Panicum joori Vasey. Along Cache River, near Heron Pond, Johnson County, by J. E. White in 1969.

Panicum mattamuskeetense Ashe. Mermet Conservation Area,

Massac County, collected by John Schwegman in 1966. This species still may be found at this station.

Panicum scoparioides Ashe. Northwest of Barrington, Lake County, collected by James Ozment in 1964.

Panicum oligosanthes Schult. var. *helleri* (Nash) Fern. North of Prairie du Rocher, Randolph County, collected by James Ozment and Wendell Crews in 1962.

Panicum commutatum Schult. var. *ashei* (Pearson) Fern. Giant City State Park, Jackson County, collected by the author in 1964.

Gymnopogon ambiguus (Michx.) BSP. Burke Branch, Pope County, collected by John Schwegman in 1966. This species continues to form a sizeable colony at Burke Branch.

Aristida desmantha Trin. and Rupr. First collected from Mason County in 1861 by M. S. Bebb and collected subsequently several times in the same vicinity by various botanists.

USEFULNESS OF GRASSES

Grasses are undoubtedly the most valuable plants to mankind. Grasses used for food by man and his domesticated animals are many. Man utilizes such grasses as barley, corn, millet, oats, rice, rye, sorghum, sugar cane, and wheat in his own diet. Many forage grasses are used for hay, silage, and pasturing. In the open expanses of the western United States, pasture grasses are referred to as range grasses. Many of these are important members of the grass flora of Illinois. Grains of many grasses are important in the diet of wildlife and fowl.

Man employs other grasses for his benefit and enjoyment. Grasses with strong rhizomes are an important tool against soil erosion. In the dunal region of Lake Michigan, certain grasses are valuable sandbinders. Lawn grasses become more and more important to modern living. Several attractive grasses are cultivated for their ornamental value.

RELATIONSHIP OF THE GRASSES

Grasses (Poaceae) and sedges (Cyperaceae) have long been placed near each other in most phylogenetic schemes. Indeed, these two groups share several similar characters – general habit, reduced flowers subtended by an assortment of scales or bracts, similar leaves, 1-seeded fruits, etc. Near to these families, Hutchinson (1959) and others have placed the rushes (Juncaceae) which, although similar in general habit and the presence of inconspicuous flowers, possess an actual perianth.

Recent evidence seems to point to a relationship of grasses, sedges, and rushes to several other families. In particular, grasses appear to be closely related to the Flagellariaceae (of Old World tropics and subtropics), the Restionaceae (of the Southern Hemisphere and Indochina), and the Centrolepidaceae (of the Southern Hemisphere). Thorne (1968) proposes that the Poaceae represent the highest development of the Order Commelinales which, in addition to the other families mentioned in the discussion above, includes the Bromeliaceae, Rapateaceae, Xyridaceae, Pontederiaceae, Phylidraceae, Commelinaceae, Mayacaceae, and Eriocaulaceae. Under Thorne's classification, the grasses appear as follows:

> Class Angiospermae
> Subclass Monocotyledoneae
> Superorder Commeliniflorae
> Order Commelinales
> Suborder Poineae
> Family Poaceae

CLASSIFICATION OF GRASSES

Many systems of classification have been proposed for the more than 600 genera and 10,000 species of grasses in the world. It is thought that perhaps the system which formed the basis of many of these proposals was presented in 1881 by the British botanist George Bentham. Bentham, basing his system of classification primarily on characters of the inflorescence and the florets, recognized thirteen tribes in two subfamilies. It was basically this grouping that Hitchcock followed in his standard books on the grasses of the United States in 1920 and 1935, except that Hitchcock added a fourteenth tribe, the Zizanieae, and made some minor internal adjustments.

The great percentage of floras written in the United States since 1920 has followed the system of classification for grasses as found in Hitchcock (1935) or its revised edition (Hitchcock, 1950). Thus many students of grasses have learned, for example, that *Bromus, Festuca, Poa, Eragrostis, Glyceria,* and *Uniola,* among others, are genera assigned to tribe Festuceae, while *Aristida, Stipa, Brachyelytrum, Muhlenbergia,* and *Agrostis,* among others, are genera placed in tribe Agrostideae. Those persons, accustomed to the "traditional" system, may be surprised or even shocked (although I hope not disheartened), to find in the present volumes that *Bromus, Festuca,* and *Poa* are still in the Festuceae, but that *Eragrostis* is removed to the Eragrosteae, *Glyceria* to the Meliceae, and *Uniola* (now the genus *Chasmanthium* for the Illinois species) to the Centotheceae. Moreover, the Agrostideae is no longer recognized; instead, *Aristida* is placed in the Aristideae, *Stipa* in the Stipeae, *Brachyelytrum* in the Brachyelytreae, *Muhlenbergia* in the Eragrosteae, and *Agrostis* in the Aveneae.

The reasoning for all of this reorganization is based on a great amount of experimental evidence which has been accumulating since the second quarter of the twentieth century. Chromosome studies began in earnest with the works of the Russian botanist A. P. Avdulov in 1928 and 1931, who correlated his results with information on the starch grains in the fruits, the make-up of the resting nucleus, and the anatomical structures of the leaf. John Reeder (1957) showed the significance of the embryo of grasses to their taxonomy, while W. V. Brown (1958) presented important data on leaf anatomy. More refined cytological techniques, along with chromatographic methods, have enabled the botanist to gain an even greater insight into the relationships of grass genera.

Much of the reorganization found in these volumes is based on a paper by Stebbins and Crampton in 1961, as modified by Gould in 1968. There is no attempt in this flora to explain in detail the more technical reasons why various genera are placed where they are; instead, references are given so that the interested reader can examine the reasoning.

Since most of the more recent taxonomic evidence is concerned with characters which are extremely difficult for a floristic worker to determine in a short time, it is not practicable to write a workable key to the tribes of genera of grasses which shows natural relationships. The key to genera which is presented later in this volume is referred to as an artificial key, or one that is based on easily observable characters without any attempt to depict natural affinities.

On the following pages are parallel lists showing the arrangement of the grass genera of Illinois as it would be treated under the Hitchcock system (Hitchcock, 1950) and the arrangement proposed by Gould (1968) and followed essentially in these volumes.

Alignment of Grass Genera of Illinois

Following Hitchcock System (1950)   *Following Gould System (1968)*

SUBFAMILY **Festucoideae**
Tribe *Bambuseae*
  Arundinaria
Tribe *Festuceae*
  Bromus
  Festuca
  Puccinellia
  Glyceria
  Poa
  Briza
  Eragrostis
  Diarrhena
  Redfieldia
  Distichlis
  Uniola
  Dactylis
  Phragmites
  Melica
  Schizachne
  Tridens
  Triplasis
Tribe *Hordeae*
  Agropyron
  Triticum
  Aegilops
  Secale
  Elymus
  Hystrix
  Hordeum
  Lolium
Tribe *Aveneae*
  Koeleria
  Sphenopholis
  Deschampsia
  Aira
  Avena
  Arrhenatherum
  Holcus
  Danthonia
Tribe *Agrostideae*
  Calamagrostis
  Ammophila
  Calamovilfa
  Agrostis
  Cinna
  Alopecurus
  Phleum
  Muhlenbergia
  Sporobolus
  Heleochloa
  Brachyelytrum
  Milium

SUBFAMILY **Festucoideae**
Tribe *Festuceae*
  Bromus
  Vulpia
  Festuca
  Lolium
  Puccinellia
  Poa
  Briza
  Dactylis
Tribe *Aveneae*
  Koeleria
  Sphenopholis
  Aira
  Deschampsia
  Avena
  Arrhenatherum
  Holcus
  Calamagrostis
  Ammophila
  Agrostis
  Cinna
  Anthoxanthum
  Hierochloë
  Phalaris
  Alopecurus
  Phleum
  Milium
  Beckmannia
Tribe *Triticeae*
  Elymus
  Sitanion
  Hordeum
  × Agrohordeum
  Agropyron
  Triticum
  Secale
Tribe *Meliceae*
  Melica
  Glyceria
  Schizachne
Tribe *Stipeae*
  Stipa
  Oryzopsis
Tribe *Brachyelytreae*
  Brachyelytrum
Tribe *Diarrheneae*
  Diarrhena
SUBFAMILY **Panicoideae**
Tribe *Paniceae*
  Digitaria
  Trichachne

SUBFAMILY **Festucoideae**
(*continued*)
Oryzopsis
Stipa
Aristida
Tribe *Zoysieae*
Zoysia
Tribe *Chlorideae*
Leptochloa
Eleusine
Dactyloctenium
Cynodon
Schedonnardus
Beckmannia
Spartina
Gymnopogon
Chloris
Bouteloua
Buchloë
Tribe *Phalarideae*
Hierochloë
Anthoxanthum
Phalaris
Tribe *Oryzeae*
Leersia
Tribe *Zizanieae*
Zizania
Zizaniopsis
SUBFAMILY **Panicoideae**
Tribe *Paniceae*
Trichachne
Digitaria
Leptoloma
Eriochloa
Paspalum
Panicum
Echinochloa
Setaria
Cenchrus
Tribe *Andropogoneae*
Miscanthus
Erianthus
Microstegium
Andropogon
Sorghum
Sorghastrum
Tribe *Tripsaceae*
Tripsacum
Zea

SUBFAMILY **Panicoideae**
(*continued*)
Leptoloma
Eriochloa
Paspalum
Panicum
Echinochloa
Setaria
Cenchrus
Tribe *Andropogoneae*
Miscanthus
Erianthus
Sorghum
Sorghastrum
Andropogon
Microstegium
Bothriochloa
Schizachyrium
Tripsacum
Zea
SUBFAMILY **Eragrostoideae**
Tribe *Eragrosteae*
Eragrostis
Tridens
Triplasis
Redfieldia
Calamovilfa
Muhlenbergia
Sporobolus
Crypsis
Tribe *Chlorideae*
Eleusine
Dactyloctenium
Leptochloa
Gymnopogon
Schedonnardus
Cynodon
Chloris
Bouteloua
Buchloë
Spartina
Tribe *Aeluropodeae*
Distichlis
Tribe *Aristideae*
Aristida
SUBFAMILY **Bambusoideae**
Tribe *Bambuseae*
Arundinaria
SUBFAMILY **Oryzoideae**
Tribe *Oryzeae*
Leersia
Zizania
Zizaniopsis
SUBFAMILY **Arundinoideae**
Tribe *Arundineae*

SUBFAMILY **Arundinoideae**
(*continued*)
Phragmites
Tribe *Centotheceae*
Chasmanthium
Tribe *Danthonieae*
Danthonia

HOW TO IDENTIFY A GRASS

Beginning on page 22 is a key for the identification of the genera of the grasses of Illinois. A botanical key is a device which, when properly employed, enables the user to identify correctly the plant in question. It is the intent of this key to use characters which are easy to observe and to avoid the more technical characters which often best show relationships.

Once the genus is ascertained by using the general key, the reader should turn to that genus and use the key provided to the species of that genus if more than one species occurs in Illinois. Of course, if the genus is recognized at sight, then the genera keys should be by-passed.

The keys in this work are dichotomous, i.e., with pairs of contrasting statements. Always begin by reading both members of the first pair of statements. By choosing that statement which best fits the specimen to be identified, the reader will be guided to the next proper pair of statements. Eventually, a name will be derived.

Illustrated Key to the GENERA of Grasses in Illinois

1. Culms woody_____80. **Arundinaria** *
1. Culms herbaceous.
    2. Spikelets enclosed by a spiny bur (**Fig. 27**)_____49. **Cenchrus** *
    2. Spikelets not enclosed by a spiny bur.
        3. Spikelets with one or more perfect florets (**Figs. 28 and 29**).
            4. Inflorescence solitary, racemose, paniculate, or spicate, but not digitate.
                5. Inflorescence spicate or spike-like, with one spike per culm_____*Group A*, p. 24
                5. Inflorescence solitary, racemose, or paniculate, but not composed of single spikes.

27. Spiny bur of *Cenchrus*.

28. Spikelet with one perfect floret.

6. Each spikelet with 2 or more perfect florets (Fig. 29).

7. Some part of the spikelet awned_____ _____Group B, p. 30

7. Spikelet without any awns_____ _____Group C, p. 32

6. Each spikelet with one perfect floret (sterile or staminate lemmas may be present, in addition [Fig. 30]).

8. Some part of the spikelet awned_____ _____Group D, p. 37

8. Spikelet without awns_____Group E, p. 42

4. Inflorescence digitate (the spikes and racemes radiating from near the same point [Fig. 31]) _____ _____Group F, p. 46

3. Spikelets unisexual (i.e., either all staminate or all pistillate) _____Group G, p. 47

29. Spikelet with more than one perfect floret.

30. Spikelet with one perfect floret.

31. Digitate inflorescence.

Group A

Inflorescence spicate or spike-like, with one spike per culm; spike-lets with one or more perfect florets.

1. Spikelets cylindrical, borne at swollen rachis joints, the entire spike-let falling at maturity; each glume with one awn and one tooth (**Fig. 32**)_____32. **Triticum**, p. 243
1. Spikelets not as above; rachis joints not swollen; glumes awned or awnless, but not with one awn and one tooth.
    2. Spikelets borne edgewise to the rachis; inner glume absent, ex-cept in the terminal spikelet (**Fig. 33**)_____4. **Lolium**, p. 101
    2. Spikelets borne flatwise to the rachis; glumes present on all spikelets.
        3. Each spikelet subtended and usually surpassed by one or more sterile bristles (not to be confused with awns) (**Fig. 34**)_____48. **Setaria** *
        3. Each spikelet not subtended by bristles.
            4. Each spikelet with two or more perfect florets (*Antho-xanthum* and *Phalaris* have three lemmas, but two of them are sterile).

32. Spikelet of *Triticum*.

34. Spikelet of *Setaria*.

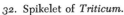

33. Inflorescence of *Lolium*.

5. At least some part of the spikelet awned.
 6. Upper spikelets paired, the lowermost solitary____
 _____30. × **Agrohordeum**, p. 228
 6. Spikelets either all paired, all borne in threes, or
 all solitary.
  7. Spikelets either all paired (**Fig. 35**) or all borne
  in threes (**Fig. 36**).
   8. Spikelets in threes_____29. **Hordeum**, p. 219
   8. Spikelets paired.
    9. Glumes to 4 cm long; axis of inflorescence
    rarely breaking apart at maturity_____
    _____27. **Elymus**, p. 201
    9. Glumes 6–8 cm long; axis of inflorescence
    breaking apart at maturity_____
    _____28. **Sitanion**, p. 217
  7. Spikelets solitary.
   10. Glumes awned.
    11. Glumes 3-nerved (**Fig. 37**)_____
    _____32. **Triticum**, p. 243
    11. Glumes 1-nerved (**Fig. 38**).

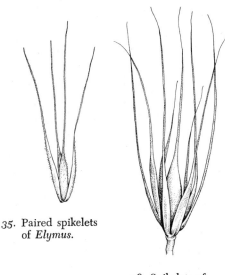

35. Paired spikelets
 of *Elymus.*

36. Spikelets of
 Hordeum.

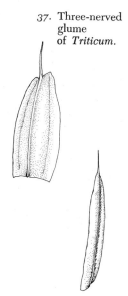

37. Three-nerved
 glume
 of *Triticum.*

38. One-nerved
 glume of *Secale.*

12. Awn of lemmas to 8 cm long; annual_____33. **Secale**, p. 244

12. Awn of lemmas to 3 cm long; perennials_____31. **Agropyron**, p. 230

10. Glumes awnless.

13. Blades 10–20 mm broad; annuals____
_____32. **Triticum**, p. 243

13. Blades 1–10 mm broad; perennials.

14. Awn of lemma more than 1 mm long_____31. **Agropyron**, p. 230

14. Awn of lemma up to 1 mm long__
_____9. **Koeleria**, p. 141

5. Spikelets awnless throughout.

15. Annuals_____32. **Triticum**, p. 243

15. Perennials.

16. Spikelets paired (**Fig. 39**)_____
_____27. **Elymus**, p. 201

16. Spikelets solitary.

17. Spikelets borne flatwise to the continuous rachis (**Fig. 40**)____31. **Agropyron**, p. 230

17. Spikelets borne all around the articulated (jointed) rachis.

18. Blades 3–10 mm broad; lemmas densely pubescent on the nerves (**Fig. 41**)_____ 61. **Tridens** °

39. Paired spikele
of *Elymus*.

41. Densely pubescent lemma of *Tridens*.

40. Rachis with spikelets in *Agroppron*.

18. Blades 1–3 mm broad; lemmas merely scabrous_____9. **Koeleria**, p. 141

4. Each spikelet with a single perfect floret (1–2 sterile lemmas present in addition in *Anthoxanthum* and *Phalaris;* 1 staminate lemma present in addition in *Holcus*).

19. Upper spikelets paired, the lowermost solitary_____ _____30. × **Agrohordeum**, p. 228

19. Spikelets either all paired, all borne in threes, or all solitary.

20. Spikelets paired or in groups of three; glumes long-awned.

21. Spikelets paired (**Fig. 42**)_____ _____27. **Elymus**, p. 201

21. Spikelets in threes (**Fig. 43**)_____ _____29. **Hordeum**, p. 219

20. Spikelets solitary.

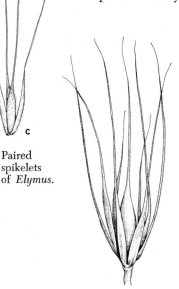

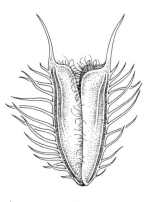

42. Paired spikelets of *Elymus*.

44. Awned glume of *Phleum*.

43. Spikelets of *Hordeum*.

22. Some part of the spikelet awned.

23. Lemmas awnless; glumes awned (**Fig. 44**)_____24. **Phleum**, p. 197

23. Lemmas awned; glumes awned or awnless.

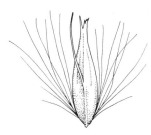

45. Lemma awned from middle in *Calamagrostis*.

46. Lemma awned from tip in *Alopecurus*.

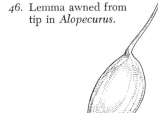

24. Lemma awned from the middle (**Fig. 45**)_____16. **Calamagrostis**, p. 158
24. Lemma awned from the tip (**Fig. 46**).
    25. Lemma 1 per spikelet, perfect.
       26. Glumes united near the base (**Fig. 47**) _____ ____23. **Alopecurus**, p. 190
       26. Glumes free at the base__ _____65. **Muhlenbergia** °
    25. Lemmas 2–3 per spikelet, but only one perfect.
       27. Spikelets 5–10 mm long, each with one perfect floret and two empty lemmas (**Fig. 48**) _____ __20. **Anthoxanthum**, p. 181
       27. Spikelets 3.5–5.0 mm long, each with one perfect floret and one staminate floret (**Fig. 49**) _____ _____15. **Holcus**, p. 158
22. Spikelets not awned.
    28. Glumes 9–15 mm long; lemma 7–14 mm long_____17. **Ammophila**, p. 165
    28. Glumes to 7 (–10 in *Phalaris*) mm long; lemmas to 7 mm long.
       29. Each spikelet with one perfect floret and 1–2 empty lemmas (**Fig. 50**)__ _____22. **Phalaris**, p. 189
       29. Each spikelet with one perfect floret only.
          30. Lemma 3-nerved (**Fig. 51**)____ _____65. **Muhlenbergia** °
          30. Lemma 1-nerved.
             31. Spikes broad, one-fourth to one-half as broad as long

47. Glumes united near base in *Alopecurus*.

(Fig. 52)____67. **Crypsis** *

31. Spikes slender, one-fifth or
   less as broad as long (**Fig.
   53**)_____66. **Sporobolus** *

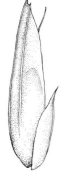

48. Spikelet of *Anthoxanthum.*

49. Spikelet of *Holcus.*

50. Spikelet of *Phalaris.*

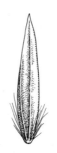

51. Lemma in
   Muhlenbergia.

52. Spike of *Crypsis.*

53. Spike of *Sporobolus.*

Group B

Inflorescence solitary, racemose, or paniculate, but not spicate or digitate; spikelets with 2 or more perfect flowers; some part of the spikelet awned.

1. Lemmas 2-toothed at the apex (**Fig. 54**).
    2. Awn of lemma arising from between the teeth (Fig. 54).
        3. Lemmas 5- to 9-nerved (**Fig. 55**).
            4. Callus of lemmas bearded (**Fig. 56**)_____ _____36. **Schizachne**, p. 265
            4. Callus of lemmas not bearded.
                5. Glumes much shorter than the entire spikelet (**Fig.** 57)_____1. **Bromus**, p. 51
                5. Glumes equalling or longer than the uppermost floret (**Fig. 58**)_____86. **Danthonia** *
        3. Lemmas 3-nerved.
            6. Panicles 3–5 (–8) cm long; spikelets 2- to 5-flowered (**Fig. 59**)_____62. **Triplasis** *

54. Apex of lemma two-toothed.

55. Lemma with an awn between the teeth.

56. Bearded callus of lemma of *Schizachne.*

57. Spikelet with short glumes in *Bromus.*

59. Spikelet of *Triplasis.*

60. Spikelet
of *Leptochloa.*

58. Spikelet with long
glumes in *Danthonia.*

6. Panicles 10–20 cm long; spikelets 6- to 12-flowered (**Fig.
60**)_____70. **Leptochloa** *
2. Awn of lemma arising near the middle or base of the lemma
(**Figs. 61 and 62**).
    7. Glumes 17–30 mm long; awn of lemmas up to 25 mm long
_____13. **Avena**, p. 152

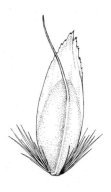

61. Awn arising from
middle of lemma
in *Aira.*

62. Awn arising from base
of lemma in *Deschampsia.*

7. Glumes 2.5–5.0 mm long; awn of lemmas 2.5–6.0 mm long.
    8. Awn arising from near the middle of the lemma; lemmas 3-nerved (Fig. 61)_____11. **Aira**, p. 147
    8. Awn arising from near base of lemma; lemmas 5-nerved (Fig. 62) _____12. **Deschampsia**, p. 151
1. Lemmas acute or obtuse at the apex, not 2-toothed.
    9. Lemmas 3-nerved_____70. **Leptochloa** °
    9. Lemmas 5-nerved (all the nerves sometimes obscure in *Festuca*).
        10. Blades involute, about 1 mm in diameter.
            11. Plants annual; stamen 1 _____2. **Vulpia**, p. 83
            11. Plants perennial; stamens 3_____3. **Festuca**, p. 88

63. Inflorescence of *Festuca*.

64. Inflorescence of *Dactylis*.

        10. Blades flat, 2–8 mm broad.
            12. Lemmas glabrous; spikelets not crowded in 1-sided panicles (**Fig. 63**)_____3. **Festuca**, p. 88
            12. Lemmas ciliate along the keel; spikelets crowded in 1-sided panicles (**Fig. 64**)_____8. **Dactylis**, p. 137

Group C

Inflorescence solitary, racemose, or paniculate, but not spicate or digitate; spikelets with 2 or more perfect flowers; spikelets awnless.

1. Lemmas distinctly 2-toothed at the apex (**Fig. 65**).
    2. Perennial; blades to 3 mm broad; panicle branches erect or spreading; spikelets 5- to 12-flowered; glumes 1 cm long____
    _____1. **Bromus**, p. 51

 2. Annual; blades 5–15 mm broad; panicle branches lax; spikelets
 2-flowered; glumes 1.5–2.5 cm long_____13. **Avena**, p. 152
1. Lemmas acute to obtuse at the apex, not 2-toothed.
 3. Glumes at least 15 mm long, as long as the spikelets (**Fig. 66**)
 _____13. **Avena**, p. 152
 3. Glumes less than 10 mm long, shorter than the spikelets.
 4. Rachilla with long silky hairs, the hairs longer than the spikelets (**Fig. 67**); culms to 4 m tall_____84. **Phragmites** *
 4. Rachilla without long silky hairs longer than the spikelets;
 culms to 1.5 m tall.
 5. Lemmas 3-nerved.
 6. Lemmas 6–10 mm long; grain beaked (**Fig. 68**)____

68. Beaked grain
 of *Diarrhena*.

67. Silky-haired
 rachilla of *Phragmites*.

65. Apex of lemma
 two-toothed. *66.* Spikelet with long
 glumes in *Avena*.

 _____40. **Diarrhena**, p. 278
 6. Lemmas 1.5–5.0 mm long; grain not beaked.
 7. Lemmas glabrous_____60. **Eragrostis** *
 7. Lemmas pubescent.
 8. Lemmas densely hairy at base, frequently with a
 tuft of hairs.
 9. Lemmas villous at the base, but without a
 tuft of cobwebby hairs.
 10. Lemmas retuse or obtuse, 3.5–4.0 mm
 long (**Fig. 69**)_____61. **Tridens** *
 10. Lemmas acute and mucronate, 4.5 mm
 long (**Fig. 70**)_____63. **Redfieldia** *
 9. Lemmas with a tuft of cobwebby hairs at the
 base, puberulent on the nerves (**Fig. 71**)__

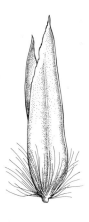

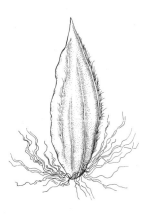

69. Acute and mucronate lemmas of *Tridens*.

70. Lemmas of *Redfieldia*.

71. Cobwebby lemma of *Poa*

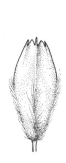

72. Spikelet of *Briza*.

         _____6. **Poa**, p. 111

  8. Lemmas pubescent only on the nerves.

      11. Lemmas keeled _____60. **Eragrostis** °

      11. Lemmas rounded on the back.

          12. Lemmas 1.0–2.5 mm long; spikelets 1–5 mm long_____70. **Leptochloa** °

          12. Lemmas 4.0 mm long; spikelets 5–8 mm long_____61. **Tridens** °

5. Lemmas 5- to many-nerved, or apparently nerveless, or with only the mid-nerve conspicuous.

  13. Lemmas apparently nerveless.

      14. Spikelets disarticulating below the glumes_____ _____10. **Sphenopholis**, p. 142

      14. Spikelets disarticulating above the glumes.

          15. Glumes 2.0–4.5 mm long; plants of moist or dry woods_____3. **Festuca**, p. 88

          15. Glumes up to 2 mm long; plants of waste ground_____5. **Puccinellia**, p. 107

  13. Lemmas obviously nerved.

      16. Lemmas 4–10 mm long.

          17. Lemmas as broad as long; inflorescence with up to eight spikelets (**Fig. 72**)_____ _____7. **Briza**, p. 135

          17. Lemmas longer than broad; inflorescence with more than eight spikelets.

18. Lemmas 4–10 mm long, with nine or more nerves.

    19. Spikelets compressed, 6- to 18-flowered (**Fig. 73**)_____ _____85. **Chasmanthium** *

    19. Spikelets not compressed, 2- to 3-flowered (**Fig. 74**)_____ _____34. **Melica**, p. 247

18. Lemmas to 7 (–8) mm long, 5- to 7-nerved.

    20. Lemmas obscurely 7-nerved (**Fig. 75**); spikelets 10–20 mm long____ _____35. **Glyceria**, p. 251

    20. Lemmas 5-nerved; spikelets less than 10 mm long.

        21. Spikelets not crowded in 1-sided panicles, not compressed (**Fig. 76**)_____3. **Festuca**, p. 88

        21. Spikelets crowded in 1-sided

73. Spikelet of
Chasmanthium.

74. Spikelet of *Melica*.

75. Spikelet of *Glyceria*.

76. Inflorescence of *Festuca*.

77. Inflorescence of *Dactylis*.

panicles, compressed (**Fig. 77**)
_ _ _ _ _ _ _ _ _ _ _8. **Dactylis**, p. 137
16. Lemmas 1.5–5.0 mm long.
   22. Lemmas distinctly keeled (**Fig. 78**)_ _ _ _ _ _
   _6. **Poa**, p. 111
   22. Lemmas rounded on the back.
      23. Nerves of lemma parallel to the summit.
         24. Sheaths closed; lodicules united
            (**Fig. 79**)_ _ _ _ _ _35. **Glyceria**, p. 251
         24. Sheaths open; lodicules free from
            each other (**Fig. 80**)_ _ _ _ _ _ _ _ _ _ _ _
            _ _ _ _ _ _ _ _ _ _ _ _5. **Puccinellia**, p. 107
      23. Nerves of lemma converging toward the
      summit.
         25. Lemmas glabrous.
            26. Plants annual; stamen 1_ _ _ _ _ _
            _ _ _ _ _ _ _ _ _ _ _ _ _2. **Vulpia**, p. 83
            26. Plants perennial; stamens 3_ _ _ _
            _ _ _ _ _ _ _ _ _ _ _ _3. **Festuca**, p. 88
         25. Lemmas pubescent, at least on the
         nerves or the keel or at the base_ _
         _ _ _ _ _ _ _ _ _ _ _ _ _ _ _ _ _ _6. **Poa**, p. 111

78. Keeled lemma of *Poa*.

79. United lodicules of *Glyceria*.

80. Free lodicules of *Puccinellia*.

Group D

Inflorescence solitary, racemose, or paniculate, but not composed of single spikes; each spikelet with one perfect floret (sterile or staminate lemmas may be present, in addition); some part of the spikelet awned.

1. Spikelets borne singly (i.e., not paired).
    2. Lemma 3-awned.
        3. Spikelets borne on one side of a long, arching raceme (**Fig. 81**); lemma rounded on the back; spikelets with one perfect lemma and 1–2 sterile ones_____75. **Bouteloua** °
        3. Spikelets borne in a more or less erect inflorescence, not 1-sided (**Fig. 82**); lemma inrolled around the palea; no sterile lemma present_____79. **Aristida** °
    2. Lemma 1-awned or awnless.
        4. First glume reduced to a sheath, united with the lowest, swollen joint of rachilla (**Fig. 83**) _____44. **Eriochloa,** p. 292

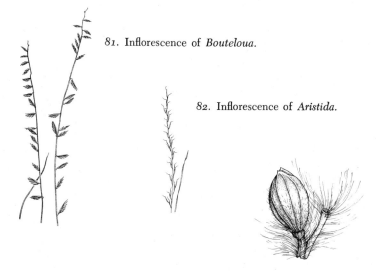

81. Inflorescence of *Bouteloua.*

82. Inflorescence of *Aristida.*

83. Spikelet of *Eriochloa.*

4. First glume not reduced to a sheath and not united with the rachillar joint.
    5. Lemma awnless; glumes awned.

6. Plants over 1 m tall; lemmas 7–10 mm long_____
_____77. **Spartina** *

6. Plants less than 1 m tall; lemmas 2–4 mm long_____
_____65. **Muhlenbergia** *

5. Lemma awned; glumes awnless or awned.

7. Spikelets arranged in 4 or more crowded ranks, each spikelet composed of one fertile and one sterile floret (**Fig. 84**)_____47. **Echinochloa** *

7. Spikelets not arranged in 4 or more crowded ranks, each spikelet composed of one fertile floret (also one staminate floret in *Arrhenatherum* or one sterile floret sometimes in *Gymnopogon*).

8. Blades 1–3 mm broad.

9. First glume less than 1 mm long_____
_____ 65. **Muhlenbergia** *

9. First glume at least 1.5 mm long.

10. Awn of lemma 2–4 cm long.

11. Tufted annual from a cluster of fibrous roots _____ 79. **Aristida** *

11. Cespitose or stout perennial_____
_____37. **Stipa,** p. 265

84. Four-ranked spikelets of Echinochloa.

10. Awn of lemma up to 2 cm long.

12. Lemma 1.0–1.6 mm long_____
_____18. **Agrostis,** p. 165

12. Lemma 2.0–4.5 mm long.

13. Glumes 5-nerved (**Fig. 85**); lemma indurated____38. **Oryzopsis,** p. 271

13. Glumes 1-nerved (**Fig. 86**); lemma not indurated__65. **Muhlenbergia** *

86. Glume of *Muhlenbergia.*

85. Spikelet of *Oryzopsis.*

8. Blades 3 mm broad or broader.
   14. Second glume 5- to 7-nerved.
      15. Awns straight or curved, not twisted near base; second glume 7-nerved (**Fig. 87**)____ _____38. **Oryzopsis**, p. 271
      15. Awns twisted near base; second glume 5-nerved (**Fig. 88**)_____37. **Stipa**, p.265
   14. Second glume 1- to 3-nerved.
      16. Lemma (excluding awns) 5–10 mm long.
         17. First glume less than 1 mm long____ _____39. **Brachyelytrum**, p. 276
         17. First glume 2.5–8.0 mm long.
            18. Awn 10–20 mm long; spikelet (excluding awns) 7–10 mm long; lemma 5- to 7-nerved_____ _____14. **Arrhenatherum**, p. 155
            18. Awn to 1.5 mm long; spikelet (excluding awns) 2.5–6.5 mm long; lemma 3-nerved _____ _____19. **Cinna**, p. 179
      16. Lemma (excluding awns) 1.5–5.0 mm long.
         19. Spikelets remote along one side of a slender rachis, forming very slender unilateral spikes (**Fig. 89**)_____ _____71. **Gymnopogon** *
         19. Spikelets in contracted or open panicles.
            20. Lemma with a tuft of hairs at the base (on the callus), awned from near the middle (**Fig. 90**)_____ _____16. **Calamagrostis**, p. 158

87. Spikelet of *Oryzopsis.*

88. Spikelet of *Stipa.*

89. Spike of *Gymnopogon.*

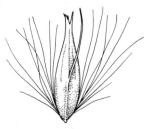

90. Lemma of *Calamagrostis.*

20. Lemma glabrous or pubescent, but without a large tuft of hairs on the callus, awned from the tip.

    21. Plants at least 1 m tall; spikelets disarticulating below the glumes _____19. **Cinna**, p. 179

    21. Plants up to 1 m tall, usually smaller; spikelets disarticulating above the glumes_____ _____65. **Muhlenbergia** °

1. Spikelets borne in pairs.

  22. Both spikelets pedicellate, the pedicels unequal in length (**Fig. 91**)_____50. **Miscanthus** °

  22. One spikelet sessile, the other pedicellate (or represented merely by the pedicel).

    23. Pedicellate spikelet represented only by the pedicel (**Fig. 92**)_____53. **Sorghastrum** °

    23. Pedicellate spikelet present.

      24. Both spikelets of the pair with perfect florets_____ _____51. **Erianthus** °

      24. Only the sessile spikelet of the pair perfect.

        25. Inflorescence racemose or nearly spicate.

92. Paired spikelets of *Sorghastrum*.

91. Paired spikelets of *Miscanthus*.

26. Flowering culms much branched into many short leafy branchlets terminated by 1–6 racemes.

    27. Racemes 2 or more from the sheaths (**Fig. 93**)_____ 54. **Andropogon** *

    27. Raceme solitary at the tip of the peduncle (**Fig. 94**)_____57. **Schizachyrium** *

26. Flowering culms unbranched_____

_____56. **Bothriochloa** *

25. Inflorescence paniculate (**Fig. 95**)_____

_____52. **Sorghum** *

93. Inflorescence of *Andropogon.*

95. Inflorescence of *Sorghum.*

94. Inflorescence of *Schizachyrium.*

Group E

Inflorescence solitary, racemose, or paniculate, but not spicate; each spikelet with one perfect floret (sterile or staminate lemmas may be present, in addition); no part of the spikelet awned.

1. Spikelets borne in pairs.
   2. One spikelet of the pair sessile, the other pedicellate (**Fig. 96**) _____55. **Microstegium** *
   2. Both spikelets either sessile or pedicellate.
      3. First glume as long as or longer than the lemmas (**Fig. 97**); plants 2.5–4.0 m tall_____50. **Miscanthus** *
      3. First glume absent or up to 0.5 mm long, much shorter than the lemmas; plants up to 1.5 m tall.
         4. Spikelets with long, tawny hairs longer than the spikelets (**Fig. 98**)_____41. **Trichachne**, p. 287
         4. Spikelets without long, tawny hairs exceeding the spikelets (**Fig. 99**)_____45. **Paspalum**, p. 296
1. Spikelets solitary (i.e., not borne in pairs).
   5. First glume reduced to a sheath and united with the lowest, swollen joint of the rachilla (**Fig. 100**)____44. **Eriochloa**, p. 292

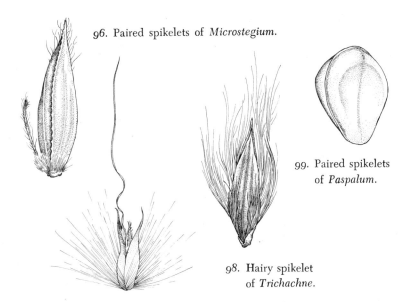

96. Paired spikelets of *Microstegium*.

99. Paired spikelets of *Paspalum*.

98. Hairy spikelet of *Trichachne*.

97. Spikelet of *Miscanthus*.

5. First glume absent, reduced, or normal, neither sheath-like nor united with a swollen rachillar joint.

    6. Both glumes absent (**Fig. 101**)_____81. **Leersia** °

    6. Both glumes present, although the first often much reduced or, if absent, the plants not producing seeds.

       7. First glume absent (**Fig. 102**); plants with creeping rhizomes, rarely producing seeds_____**Zoysia** [1]

       7. First glume present, although occasionally strongly reduced; rhizome present or absent; plants producing seeds.

          8. First glume up to one-half (to ⅔ in a few species of *Panicum*) as long as second glume.

---

[1] **Zoysia** is frequently planted in Illinois as a choice lawn grass, but no collections have ever been made of it as an adventive. Therefore, it is excluded from the text.

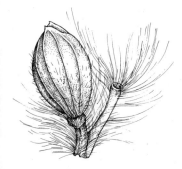

100. Spikelet of *Eriochloa*.

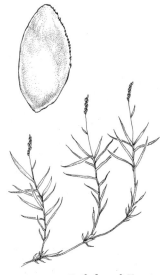

102. Spikelet of *Zoysia*.

101. Spikelet of *Leersia*.

9. Each floret subtended by one or more bristles (**Fig. 103**)_____481. **Setaria** *
9. Florets not subtended by bristles.
   10. Spikelets solitary at the end of long, capillary pedicels.
      11. Fertile lemma leathery_____ _____43. **Leptoloma**, p. 289
      11. Fertile lemma indurated____46. **Panicum** *
   10. Spikelets grouped in 2–4 or more ranks.
      12. Inflorescence a dense, dark purple-brown panicle_____47. **Echinochloa** *
      12. Inflorescence racemose or paniculate with remote, ascending racemes.
         13. Racemes 1–2 (–3) cm long_____ _____47. **Echinochloa** *
         13. Racemes over 2 cm long_____ _____45. **Paspalum**, p. 296
8. First glume nearly as long as the second glume, not conspicuously different in size.
14. Glumes 3-nerved (**Fig. 104**).
   15. Spikelets 4.5–6.0 mm long; glumes 4–6 mm long; blades 2–5 mm broad_____ _____21. **Hierochloë**, p. 186
   15. Spikelets 2.0–3.5 mm long; glumes 2–3 mm long; some or all the blades over 5 mm broad.
      16. Lemmas 5-nerved; spikelet with one perfect and one sterile lemma (**Fig. 105**); blades to 8 mm broad_____ _____26. **Beckmannia**, p. 198

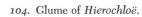
104. Glume of *Hierochloë.*

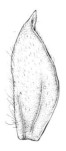
103. Floret in *Setaria* with bristles.

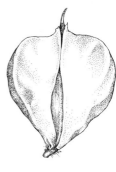
105. Spikelet of *Beckmannia.*

16. Lemma nerveless; spikelet with one perfect lemma (**Fig. 106**); blades to 20 mm
broad_____25. **Milium**, p. 197

14. Glumes 1-nerved.

17. Lemma with a conspicuous tuft of hairs at the
base (on the callus) (**Fig. 107**); spikelets 6–7
mm long_____64. **Calamovilfa** *

17. Lemma glabrous or pubescent, but without a
large tuft of hairs on the callus; spikelets 1–6
mm long.

18. Lemma 3- to 5-nerved, the nerves sometimes obscure.

19. First glume longer than the lemma
(**Fig. 108**)_____18. **Agrostis**, p. 165

19. First glume shorter than the lemma.

20. Spikelets appressed on two sides
of a triangular rachis (**Fig. 109**)
_____72. **Schedonnardus** *

20. Spikelets not confined to two sides
of the rachis_____
_____65. **Muhlenbergia** *

18. Lemma 1-nerved_____66. **Sporobolus** *

106. Spikelet of Milium.

108. Spikelet of Agrostis.

107. Lemma of Calamovilfa.

109. Spikelets of Schedonnardus.

Group F

Inflorescence digitate (the spikes and racemes radiating from near the same point).

1. Some part of the spikelet awned.
    2. Spikelets borne in pairs, one sessile and perfect, the other pedicellate and staminate_____54. **Andropogon** *
    2. Spikelets borne singly.
        3. Spikelets with 3–5 perfect florets; second glume and lemmas awned (**Fig. 110**)_____69. **Dactyloctenium** *
        3. Spikelets with 1 perfect floret (also 1–2 empty lemmas present); fertile lemma awned (**Fig. 111**)_____74. **Chloris** *

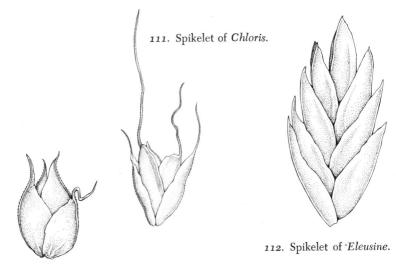

111. Spikelet of *Chloris.*

112. Spikelet of *Eleusine.*

110. Spikelet of *Dactyloctenium.*

1. Spikelets awnless.
    4. Spikelets with 3–6 perfect florets (**Fig. 112**)____68. **Eleusine** *
    4. Spikelets with 1 perfect floret.
        5. First glume 1.0–1.5 mm long; second glume 1-nerved; no sterile lemmas present_____73. **Cynodon** *
        5. First glume absent or up to 0.8 mm long; second glume 5-nerved; lower lemma empty_____41. **Digitaria,** p. 280

Group G

Spikelets unisexual (i.e., either all staminate or all pistillate).

1. Plants to 40 cm tall, dioecious; staminate spikelets 3- to 75-flowered.
    2. Lemmas with a tuft of cobwebby hairs at base (**Fig. 113**)_____
    _____6. **Poa**, p. 111

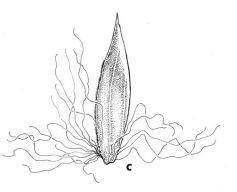

113. Cobwebby lemma in *Poa.*

114. Pistillate and staminate spikelets of *Eragrostis reptans.*

115. Pistillate and staminate spikelets of *Buchloë.*

    2. Lemmas without a tuft of cobwebby hairs at base.
        3. Both staminate and pistillate spikelets 10- to 75-flowered (**Fig. 114**)_____60. **Eragrostis** *
        3. Staminate spikelets 3- to 15-flowered, pistillate spikelets 1- to 9-flowered.
            4. Staminate spikelets 3-flowered; pistillate spikelets 1-flowered (**Fig. 115**)_____76. **Buchloë** *

4. Staminate spikelets 8- to 15-flowered; pistillate spikelets 7- to 9-flowered (**Fig. 116**)_____78. **Distichlis** *

1. Plants 1–4 m tall, monoecious; staminate spikelets 1- to 2-flowered.
    5. Staminate spikelets 2-flowered; glumes membranous.
        6. Annual; staminate and pistillate spikelets in different inflores-
        cences; pistillate spikelets borne in pairs (**Fig. 117**)_____
        _____59. **Zea** *
        6. Perennial; staminate and pistillate spikelets in the same in-
        florescence; pistillate spikelets solitary (**Fig. 118**)_____
        _____58. **Tripsacum** *

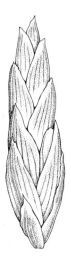

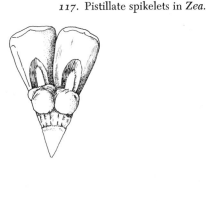

117. Pistillate spikelets in *Zea.*

116. Pistillate and staminate spikelets of *Distichlis.*

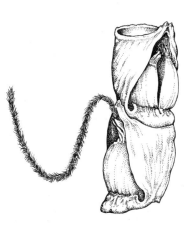

118. Pistillate spikelets in *Tripsacum.*

5. Staminate spikelets 1-flowered; glumes none.

    7. Pistillate spikelets confined to the uppermost erect branches of the inflorescence, the staminate spikelets confined to the lower spreading branches (**Fig. 119**); margin of leaf more or less smooth_____82. **Zizania** °

    7. Pistillate and staminate spikelets on the same branches of the inflorescence (**Fig. 120**); margin of leaf harsh and cutting__ _____83. **Zizaniopsis** °

119. Inflorescence in *Zizania.*

120. Inflorescence in *Zizaniopsis.*

Descriptions and Illustrations

Order Commelinales

POACEAE – GRASS FAMILY

Annual or perennial herbs (woody in the Bambuseae); culms cylindrical, with usually hollow internodes and closed nodes; leaves alternate, 2-ranked; sheaths usually free; ligule mostly present; inflorescence composed of (1–) several spikelets; spikelets 1- to several-flowered, each with usually a pair of sterile scales (glumes) at the base; flowers usually perfect, without a true perianth, the perianth reduced to rudiments (lodicules) or absent; flowers subtended by a lemma and a palea; stamens (1–) 3 (–6); ovary 1-celled, with 1 ovule; stigmas 2 (–3); fruit usually a caryopsis (grain).

This family is frequently known as the Gramineae. It is one of the largest and economically most important families of flowering plants in the world.

In the system of classification followed in this treatment, the Poaceae are one of six families comprising the order Commelinales in Illinois.

Four subfamilies of grasses occur in Illinois.

SUBFAMILY **Festucoideae**

Annuals or perennials; ligules membranous; spikelets 1- to several-flowered, sometimes with sterile florets above the fertile; spikelets disarticulating above the glumes; lemmas mostly 5- to several-nerved.

Illinois species of subfamily Festucoideae fall into seven tribes —Festuceae, Aveneae, Triticeae, Meliceae, Stipeae, Brachyelytreae, and Diarrheneae.

Subfamily Festucoideae is similar to subfamily Eragrostoideae in characters of the spikelet, although the number of nerves of the lemma is generally five or more in the Festucoideae and generally 1 or 3 in the Eragrostoideae. The major differences lie in the so-called "new" taxonomic criteria—epidermal cells in the region of root hairs, types of stomata, leaf anatomy, vascular traces in the embryo, basic chromosome number, type of reserve food, and characters concerning seed germination. Because many of these criteria are characters which are not readily observable, it is not possible to write a practical key to the tribes of subfamily Festucoideae.

Tribe *Festuceae*

Annuals or mostly perennials; inflorescence paniculate, less commonly racemose, rarely spicate; spikelets 2- to several-flowered; glumes unequal, shorter than the lowest lemma; lemmas 5- to several-nerved, awned or awnless.

There are eight genera of tribe Festuceae in Illinois, of which *Bromus, Poa,* and *Festuca* are the largest and most important.

1. *Bromus* L. – Brome Grass

Annuals or perennials from fibrous roots or rhizomes; blades usually flat, with closed sheaths; inflorescence paniculate or racemose; spikelets many-flowered, disarticulating above the glumes; glumes 2, unequal, shorter than the spikelets, 1- to several-nerved, awnless; lemmas rounded or keeled on the back, 5- to several-nerved, 2-toothed at the apex, usually with a terminal awn from between the teeth; palea shorter than the lemma, keeled, ciliate.

Some of our species are native woodland species, although most are introductions. Some of the introduced species are widespread weeds.

In the western United States, *B. ciliatus, B. inermis,* and *B. mollis* are forage grasses, while *B. willdenovii* is used for forage in the southern United States. Most of the annual weedy species are known as cheat, since they infest wheat fields and diminish the production of wheat. *Bromus tectorum, B. sterilis,* and *B. rigidus* may cause death to cattle if eaten.

The species of *Bromus* in Illinois fall into four sections. Section Pnigma contains all of the native species, along with adventives *B. erectus* and *B. inermis.* This section is characterized by species which have lemmas unkeeled and which have the teeth of the bifid apex less than 0.5 mm long. A similar group with very short lemma-teeth, but with a strong keel on the lemma, is Section Ceratochloa. This section includes *B. marginatus* and *B. willdenovii.*

There are two sections of adventive *Bromus* with the lemma-teeth greater than 0.5 mm in length. Section Zeobromus, with elliptic or oblong grains and with the first glume three-nerved, is composed of *B. sterilis* and *B. tectorum* in Illinois.

Many of the species of *Bromus* are questionably distinct. These will be discussed under the individual species.

Species 4, 5, 14, 15, 17, and 18, which belong to Section Pnigma (= § Bromopsis, = § Zerna), have been studied by Wagnon (1952). An earlier work of the genus, covering all North American species north of Mexico, is by Shear (1900).

KEY TO THE SPECIES OF Bromus IN ILLINOIS

1. Some or all of the awns over 12 mm long; teeth of lemmas 2–5 mm long.
   2. Lemmas 16–21 mm long, scabrous or puberulent on the back; first glume 8–12 mm long; second glume 13–18 mm long; awns 20–30 mm long; blades and sheaths glabrous or short-pubescent _____1. *B. sterilis*
   2. Lemmas 10–12 mm long, villous throughout on the back, becoming hispidulous at the summit, rarely entirely glabrous; first glume 4–7 mm long; second glume 8–10 mm long; awns (10–) 12–15 mm long; blades and sheaths soft-pubescent_____ _____2. *B. tectorum*
1. All or most of the awns less than 12 mm long; teeth of lemmas usually less than 2 mm long.
   3. First glume 3- to 5-nerved (1-nerved in *B. nottowayanus*); second glume 5- to 7-nerved; annuals or perennials.
      4. Perennials from rhizomes; blades 6–13 mm broad (occasionally 5 mm broad in *B. kalmii*).
         5. Lemmas keeled; inflorescence erect____3. *B. marginatus*
         5. Lemmas rounded on the back; inflorescence drooping.
            6. Awns 5–8 mm long; cauline leaves 6–8_____ _____4. *B. nottowayanus*
            6. Awns 1–3 mm long; cauline leaves 3–5 (–6) _____ _____5. *B. kalmii*
      4. Annuals from fibrous roots; blades 2–6 mm broad (to 8 mm broad only in *B. secalinus*).
         7. Lemmas strongly keeled on the back, 12–15 mm long____ _____6. *B. willdenovii*
         7. Lemmas rounded on the back, 5–11 (–12) mm long.
            8. Awns 0–6 mm long.
               9. Blades harshly pubescent above, or glabrous; lemmas 5–8 mm long; awns 1–6 mm long, rarely ab-

sent_____7. *B. secalinus*
    9. Blades softly villous; lemmas 9–11 (–12) mm long;
       awns 0–1 mm long_____8. *B. brizaeformis*
  8. Most or all the awns over 6 mm long.
    10. Inflorescence erect or ascending.
      11. Lemmas plicate, conspicuously nerved; inflo-
          rescence compact_____9. *B. mollis*
      11. Lemmas not plicate, faintly nerved; inflores-
          cence open.
        12. Lower lemmas 7–9 mm long; branches of
            inflorescence solitary or paired, usually
            shorter than the spikelets; anthers 2.0–2.5
            mm long_____10. *B. racemosus*
        12. Lower lemmas 9–11 mm long; branches of
            inflorescence 2–6, usually much longer
            than the spikelets; anthers 1.5–2.0 mm
            long_____11. *B. commutatus*
    10. Inflorescence spreading or drooping.
      13. Awn straight or nearly so; rachilla not exposed
        at maturity.
        14. Lemmas all nearly the same length; an-
            thers 4 mm long_____12. *B. arvensis*
        14. Lowest lemmas longer than the upper; an-
            thers 2.0–2.5 mm long__10. *B. racemosus*
      13. Awn flexuous; rachilla exposed at maturity___
        _____13. *B. japonicus*
3. First glume 1-nerved; second glume 3- to 5-nerved; perennials.
  (*Bromus nottowayanus* has the first glume 1-nerved, the second
  glume 5- to 7-nerved).
    15. Awns absent or up to 2 mm long; blades and sheaths gla-
      brous_____14. *B. inermis*
    15. Awns 2–8 mm long; blades and sheaths (particularly the
      lower) pubescent, rarely glabrous.
      16. Inflorescence narrow, erect; blades 2–3 mm broad____
        _____15. *B. erectus*
      16. Inflorescence spreading or drooping; blades (3–) 4–17
        mm broad.
        17. Leaves 10–20 per culm, the blades auriculate at
          base_____16. *B. purgans*
        17. Leaves 5–8 per culm, the blades not auriculate.
          18. Lemmas pubescent throughout on the back, or
            glabrous_____17. *B. pubescens*

18. Lemmas pubescent only on the margins in the
lower one-half to three-fourths of the lemma__
_____18. *B. ciliatus*

1. **Bromum sterilis** L. Sp. Pl. 77. 1753. *Fig. 121.*

Annual with glabrous culms to nearly 1 m tall; sheaths glabrous
or pubescent; blades glabrous or pubescent, 2–4 mm broad; inflorescence nodding, 10–25 cm long; spikelets (including awns)
3.5–5.0 cm long, flattened, 5- to 10-flowered; glumes glabrous or
puberulent, the first 1-nerved, 8–12 mm long, subulate, the second
3-nerved, 13–18 mm long, broader; lemmas 7-nerved, 16–21 mm
long, scabrous or puberulent; awns 20–30 mm long; anthers 1.0–
1.5 mm long.

COMMON NAME: Brome Grass.
HABITAT: Waste ground.
RANGE: Native of Europe; occasionally escaped through-
out the United States.
ILLINOIS DISTRIBUTION: Collected only twice as an escape
(Cook Co.: Chicago, June 1905, *F. C. Gates 447;* Cham-
paign Co.: Urbana, May 22, 1941, *G. N. Jones 13888*).
This *Bromus* has the longest awns of any species of
Bromus in Illinois. The only Illinois collections seen
were taken in flower during May or June.

This and the following species comprise Section Stenobromus,
characterized by being annuals with compressed spikelets, long-
awned lemmas, and lemmas with the teeth 2–5 mm long. *Bromus
sterilis* differs from *B. tectorum* by its longer spikelets and longer
awns.

The Gates collection in June, 1905, from Chicago, reported by
Jones and Fuller (1955) as *Bromus rigidus* L., is actually *B.
sterilis.*

2. **Bromus tectorum** L. Sp. Pl. 77. 1753. *Fig. 122.*

Annual with culms to about 75 cm tall; sheaths soft-pubescent;
blades soft-pubescent, 2–4 broad; inflorescence drooping, 5–20
cm long; spikelets (including awns) 2.0–3.5 cm long, flattened,
5- to 12-flowered; glumes sparsely pilose, villous, or glabrous, the
first 1-nerved, 4–7 mm long, subulate, the second 3-nerved, 8–10
mm long, narrowly lanceolate; lemmas 5- to 7-nerved, 10–12 mm
long, villous throughout on the back, becoming hispidulous near

121. *Bromus sterilis* (Brome Grass). *a.* Inflorescence, X½. *b.* Sheath, X7½. *c.* Spikelet, X1½. *d.* First glume, X5. *e.* Second glume, X5. *f.* Lemma, X5.

122. *Bromus tectorum* (Downy Chess). *a.* Inflorescence, X½. *b.* Sheath, with ligule, X5. *c.* Spikelet, X2½. *d.* First glume, X5. *e.* Second glume, X5. *f.* Lemma, X5.

the apex; awns (10–) 12–15 mm long; anthers 0.7–1.0 mm long; 2n = 14 (Knowles, 1944).

COMMON NAME: Downy Chess; Hairy Chess.
HABITAT: Waste areas.
RANGE: Native of Europe; introduced throughout the United States.
ILLINOIS DISTRIBUTION: Common; in every county.

The common name is derived from the soft pubescence of the plant. Rarely specimens with glabrous lemmas may be encountered. These are known as var. *glabratus*. A single specimen of this variety from St. Clair County has been seen. Apparently the first collection of *B. tectorum* in Illinois was in 1889 from Hilton, Tazewell County, by McDonald.

Downy Chess flowers from May to mid-August. The spikelets sometimes turn purplish at maturity.

3. **Bromus marginatus** Nees ex Steud. Syn. Pl. Glum. 1:322. 1854. *Fig. 123.*

Perennial with harshly pubescent culms to nearly 1 m tall; sheaths retrorsely and harshly pilose to glabrous; blades harshly pubescent on both surfaces, 6–12 mm broad; inflorescence erect, paniculate, 10–30 cm long; spikelets 2.5–4.0 cm long, compressed, 5- to 10-flowered; glumes glabrous, scabrous or puberulent, the first 3- to 5-nerved, 7–10 mm long, the second 5- to 7-nerved, 9–12 mm long; lemmas 7- to 9-nerved, 12–15 mm long, keeled, pubescent; awns 4–7 mm long; 2n = 56 (Stebbins & Tobgy, 1944).

COMMON NAME: Brome Grass.
HABITAT: Waste ground.
RANGE: Native of the western United States; occasionally adventive in the eastern United States.
ILLINOIS DISTRIBUTION: Collections from Cook and Kane counties have been seen.

There is disagreement as to the distinctness of this species from *B. willdenovii. Bromus marginatus* seemingly has more erect panicles, generally more pilose sheaths and culms, and longer awns. These two species are the only representatives in Illinois of Section Ceratochloa.

Although the first reference to this species as being in Illinois

123. Bromus marginatus (Brome Grass). *a.* Inflorescences, X½. *b.* Sheath, with ligule, X5. *c.* Spikelet, X2½. *d.* First glume, X5. *e.* Second glume, X5. *f.* Lemma, X5.

was made in 1950 by Hitchcock, the only specimens I have seen in Illinois were collected after that time.

4. **Bromus nottowayanus** Fern. Rhodora 43:530. 1941. *Fig. 124.*

Perennial with culms to 1.2 m tall; sheaths retrorsely pilose, rarely glabrous; blades pilose, 6–12 mm broad; inflorescence drooping, 5–23 cm long; spikelets 2.0–3.5 cm long, 3- to 12-flowered; glumes densely appressed-pilose, the first 1- (rarely 3-) nerved, 5–8 mm long, the second 5- to 7-nerved, 7–10 mm long; lemmas 7- to 9-nerved, acute, 8–13 mm long, densely appressed-pilose throughout; awns 5–8 mm long; 2n = 14 (Wagnon, 1952).

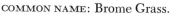

COMMON NAME: Brome Grass.

HABITAT: Moist, wooded ravines.

RANGE: Maryland to Illinois and Missouri, south to Texas and North Carolina.

ILLINOIS DISTRIBUTION: Rare; recorded from Cook, Stark, Woodford, and Peoria counties.

This is an enigmatic species with a rather disjunct distribution. It resembles *B. purgans* but lacks the great number of leaves of that species; it resembles *B. kalmii* but has longer awns, longer spikelets, and is usually more pubescent. This is the only species of *Bromus* in which the first glume is 1-nerved and the second glume is 5- to 7-nerved. It belongs to Section Pnigma.

The first collection from Illinois, made by Virginius Chase from Stark County in 1900, was originally identified as *B. purgans* var. *incanus*.

5. **Bromus kalmii** Gray, Man. 600. 1848. *Fig. 125.*

Perennial with culms to 1 m tall, pubescent or glabrous at the nodes; lower sheaths pilose, the upper glabrous; leaves glabrous, sparsely pilose, or villous, 4–10 mm broad; inflorescence drooping, 5–14 cm long; spikelets 1.5–2.5 cm long, 7- to 11-flowered; glumes pilose, the first 3-nerved, 5–7 mm long, the second 5-nerved, 6.5–8.0 mm long; lemmas 7-nerved, 7–10 mm long, villous throughout; awns 1–3 mm long; 2n = 14 (Wagnon, 1952).

124. Bromus nottowayanus (Brome Grass). *a.* Inflorescence, X½. *b.* Sheath, with ligule, X5. *c.* Spikelet, X2½. *d.* Lemma, X5.

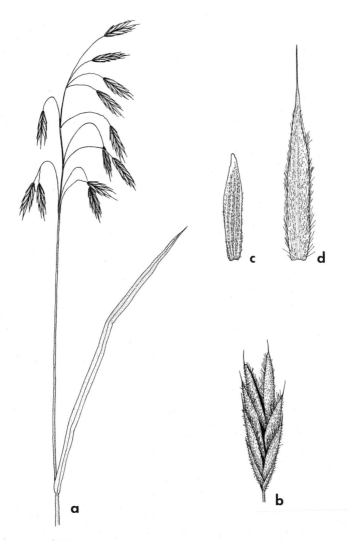

125. *Bromus kalmii* (Brome Grass). *a.* Inflorescence, X½. *b.* Spikelet, X2. *c.* First glume, X4. *d.* Lemma, X4.

COMMON NAME: Brome Grass; Wild Chess.

HABITAT: Prairie remnants and moist, calcareous fens.

RANGE: Maine to Manitoba, south to North Dakota, Illinois, and Maryland.

ILLINOIS DISTRIBUTION: Not common; confined to the northern half of the state. The specimen collected by Lapham and attributed to Jackson County by Mosher (1918) could not be located during this study.

This species flowers during June and July. Variation exists in the pubescence of the sheath and blades.

6. **Bromus willdenovii** Kunth, Rev. Gram. 1:134. 1829. *Fig. 126.*

Bromus catharticus Vahl, Symb. Bot. 2:22. 1791, nomen illeg. Usually an annual with glabrous culms to 80 cm tall; lower sheaths usually villous, the upper usually glabrous; blades pilose, 4–6 mm broad; inflorescence spreading, 5–20 cm long; spikelets 2.0–3.5 cm long, 3- to 12-flowered, flattened; glumes glabrous or scabrous, the first 3-nerved, 8–10 mm long, the second 5-nerved, 9–11 mm long; lemmas 7- to 9-nerved, 12–15 mm long, glabrous, scabrous, or puberulent, keeled; awn absent, or up to 5 mm long; 2n = 42 (Brown, 1950).

COMMON NAME: Rescue Grass.

HABITAT: Waste ground.

RANGE: Native of tropical America; introduced in most parts of the United States.

ILLINOIS DISTRIBUTION: Only a single collection seen (Champaign Co.: along railroad, Urbana, May 27, 1953, *Ahles 7370*).

This species, along with *B. marginatus*, comprises Section Ceratochloa in Illinois. The section is characterized by the flattened spikelets and the keeled lemmas. *Bromus willdenovii* differs from *B. marginatus* in its annual habit, its more glabrous herbage, its spreading inflorescence, and its generally shorter awns.

This species is more commonly known as *B. catharticus*, but this latter binomial is not valid.

7. **Bromus secalinus** L. Sp. Pl. 76. 1753. *Fig. 127.*

Annual with glabrous culms to nearly 1 m tall; lower sheaths puberulent to glabrous above, puberulent or glabrous below, 3–8

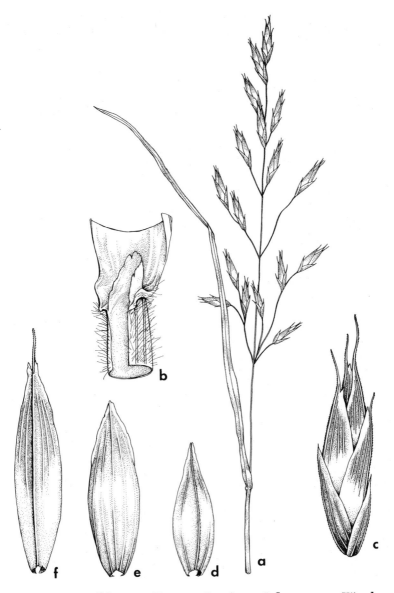

126. *Bromus willdenovii* (Rescue Grass). *a.* Inflorescence, X½. *b.* Sheath, with ligule, X5. *c.* Spikelet, X2½. *d.* First glume, X3½. *e.* Second glume, X3½. *f.* Lemma, X3½.

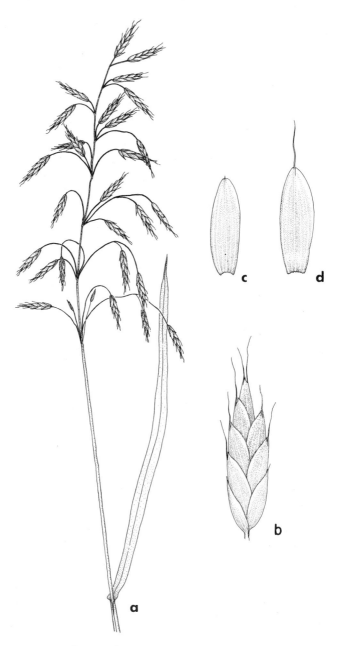

127. *Bromus secalinus* (Chess). *a.* Inflorescence, X½. *b.* Spikelet, X2½.
c. Second glume, X4. *d.* Lemma, X4.

mm broad; inflorescence nodding, 7–17 cm long; spikelets 1–2 cm long, 5- to 15-flowered; glumes glabrous, the first oblong, 3- to 5-nerved, 6–8 mm long, elliptic, obtuse, involute at the apex, glabrous or scabrous; awn (1–) 3–6 mm long, rarely absent; 2n = 28 (Knowles, 1944).

COMMON NAME: Chess; Cheat.

HABITAT: Waste ground, fields.

RANGE: Native of Europe; established in nearly all parts of the United States.

ILLINOIS DISTRIBUTION: Common; probably in most or all of the Illinois counties.

Variation in length of the awns exists in Illinois specimens; rarely, the awn is completely lacking, or reduced to 1 mm long. This species flowers from May to August.

It is one of eight introduced species from Section Zeobromus in Illinois, and may become a serious pest in grain fields.

8. **Bromus brizaeformis** Fisch. & Mey. Ind. Sem. Hort. Petrop. 3:30. 1837. *Fig. 128.*

Annual with glabrous culms to 75 cm tall; sheaths pilose to softly villous; blades pilose to softly villous on both sides, 3–6 mm broad; inflorescence drooping, 5–15 cm long; spikelets 1.0–2.5 cm long, 8- to 15-flowered, flattened; glumes obtuse, glabrous, the first 3- to 5-nerved, 5–8 mm long, the second 5- to 9-nerved, 6–9 mm long; lemmas 7- to 9-nerved, 9–12 mm long, obtuse, glabrous or puberulent; awns absent or to 1 mm long; 2n = 14 (Avdulov, 1928).

COMMON NAME: Rattlesnake Chess; Quake Grass.

HABITAT: Waste ground.

RANGE: Native of Europe; found in nearly all parts of the United States.

ILLINOIS DISTRIBUTION: Rare; two collections seen from southern Illinois.

This nearly awnless species apparently has not been collected in Illinois since 1902. Its broad, flattened spikelets are somewhat reminiscent of the genus *Briza*. The flowers appear during June and July. Many years ago this species was a favorite among ornamental grasses because of its slender, drooping panicles.

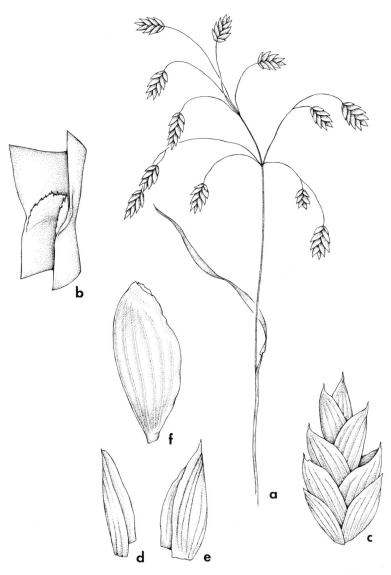

128. *Bromus brizaeformis* (Rattlesnake Chess). *a.* Inflorescence, X½. *b.* Sheath, with ligule, X5. *c.* Spikelet, X2½. *d.* First glume, X5. *e.* Second glume, X5. *f.* Lemma, X5.

9. **Bromus mollis** L. Sp. Pl., ed. 2, 112. 1762. *Fig. 129.*

Annual with culms to 80 cm tall; sheaths and blades softly pubescent, the blades 3–6 mm broad; inflorescence erect, contracted, 5–10 cm long; spikelets 1.5–3.0 cm long, 6- to 10-flowered; glumes rather broad, obtuse, pilose or scabrous, the first 3- to 5-nerved, 4–6 mm long, the second 5- to 7-nerved, 7–8 mm long; lemmas 7- to 9-nerved, 7–9 mm long, obtuse, pilose or scabrous; awns 6–10 mm long; 2n = 28 (Knowles, 1944).

COMMON NAME: Soft Chess.

HABITAT: Waste ground.

RANGE: Native of Europe; occasionally established in the northern half of the United States.

ILLINOIS DISTRIBUTION: Not common; known from two counties. The DuPage County record is based on an immature specimen too young to determine accurately. The exceptionally softly pubescent sheaths and blades and the contracted, erect inflorescence aid in the identification of this species. It flowers from May to July.

10. **Bromus racemosus** L. Sp. Pl., ed. 2, 114. 1762. *Fig. 130.*

Annual with culms to 80 cm tall; sheaths retrorsely villous; blades pubescent on both surfaces, 3–6 mm broad; inflorescence nodding or somewhat ascending, 5–20 cm long; spikelets 1–2 cm long, 5- to 10-flowered; glumes glabrous, the first 3-nerved, 4.5–6.5 mm long, the second 5-nerved, 5.5–9.0 mm long; lemmas obscurely 7-nerved, 7–10 mm long, the lowermost much longer than the upper ones, glabrous or scabrous; awns 4–10 mm long; 2n = 28 (Knowles, 1944).

COMMON NAME: Chess.

HABITAT: Waste ground, fields.

RANGE: Native of Europe; occasionally introduced in the United States.

ILLINOIS DISTRIBUTION: Occasional throughout the state. A similar species is *B. mollis,* a plant with erect, contracted inflorescences. It is also extremely difficult to distinguish from *B. commutatus,* a species with longer lemmas and a more branched inflorescence. *Bromus racemosus* flowers from late May to early August.

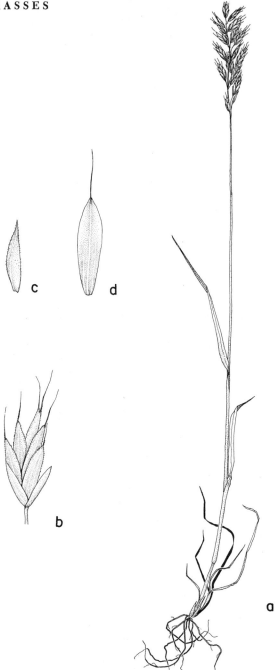

129. *Bromus mollis* (Soft Chess). *a.* Habit, X½. *b.* Spikelet, X2. *c.* First glume, X4. *d.* Lemma, X4.

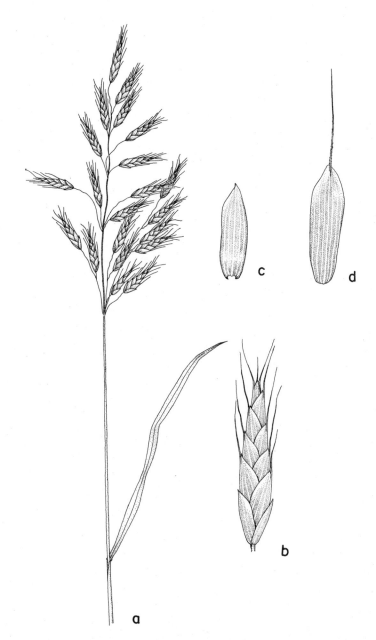

130. Bromus racemosus (Chess). *a.* Inflorescence, X½. *b.* Spikelet, X2. *c.* Second glume, X4. *d.* Lemma, X4.

11. Bromus commutatus Schrad. Fl. Germ. 353. 1806. *Fig. 131.*

Annual with culms to 75 cm tall; sheaths retrorsely pilose; blades more or less pubescent on both surfaces, 3–6 mm broad; inflorescence erect, rather open, 5–20 cm long; spikelets 1–2 cm long, 5- to 10-flowered; glumes glabrous or nearly so, the first 3-nerved, 4–6 mm long, the second 5-nerved, 6–8 mm long; lemmas obscurely 7-nerved, the lowest 9–11 mm long, glabrous or scabrous; awns 6–10 mm long; 2n = 56 (Nielsen, 1939).

COMMON NAME: Hairy Chess.

HABITAT: Waste ground, fields.

RANGE: Native of Europe; throughout most of the United States.

ILLINOIS DISTRIBUTION: Rather common throughout the state. Apparently first collected in Illinois near the turn of the century.

This species is rather questionably distinct from *B. racemosus*. The lowest lemmas in *B. commutatus* are usually larger than those in *B. racemosus*.

Bromus commutatus flowers from mid-May to early August.

12. Bromus arvensis L. Sp. Pl. 77. 1753. *Fig. 132.*

Annual with culms to 85 cm tall; sheaths softly pubescent to nearly glabrous; blades softly pubescent to nearly glabrous on both sides, 2–4 mm broad; inflorescence drooping, 10–30 cm long; spikelets 1.5–3.0 cm long, 5- to 12-flowered; glumes glabrous, the first acute, 3-nerved, 4–6 mm long, the second obtuse, 5-nerved, 5–8 mm long; lemmas obscurely 7-nerved, 7–9 mm long, obtuse, glabrous or scabrous; awns 7–10 mm long, straight; 2n = 14 (Cugnac & Simonet, 1941).

COMMON NAME: Chess.

HABITAT: Waste ground.

RANGE: Native of Europe; infrequently established in the United States, but apparently becoming more abundant.

ILLINOIS DISTRIBUTION: Not common; known only from a few counties. The first collection from Illinois of this species was made in 1941 from Jersey County by *G. D. Fuller (570)*.

This species is very similar to *B. japonicus,* but tends to have straight awns and paleas equal to the lemmas.

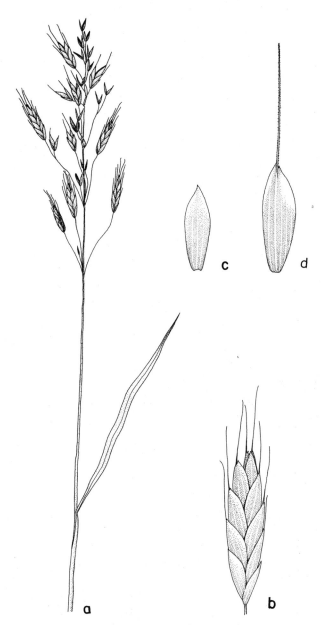

131. Bromus commutatus (Hairy Chess). *a.* Inflorescence, X½. *b.* Spikelet, X2. *c.* First glume, X3½. *d.* Lemma, X3½.

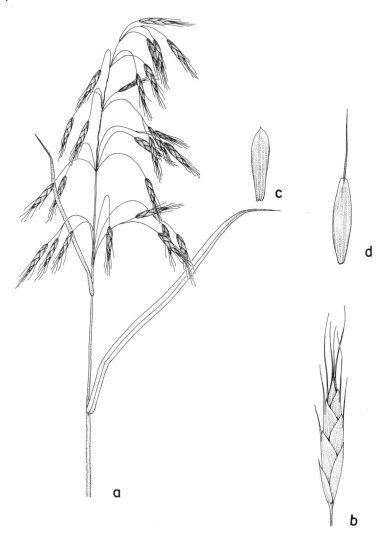

132. Bromus arvensis (Chess). *a.* Inflorescence, X½. *b.* Spikelet, X2. *c.* Second glume, X3. *d.* Lemma, X3.

Variation exists in the amount of pubescence on the sheaths and blades and in the length of the drooping inflorescence.

In Illinois, *Bromus arvensis* flowers from late May to late July. The spikelets have a tendency to become purplish at maturity.

13. Bromus japonicus Thunb. Fl. Jap. 52. 1784. *Fig. 133.*

Annual with culms to 90 cm tall; sheaths densely villous to pilose; blades densely villous to pilose on both surfaces, 2–4 mm broad; inflorescence drooping, 10–20 cm long; spikelets 2.0–2.5 cm long, 7- to 10-flowered; glumes glabrous, the first acute, 3-nerved, 4–6 mm long, the second obtuse, 5-nerved, 5.0–7.5 mm long; lemmas 7- to 9-nerved, 6.5–9.0 mm long, obtuse, glabrous; awns 5–10 (–12) mm long, flexuous.

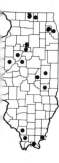

COMMON NAME: Japanese Chess.

HABITAT: Waste ground, fields.

RANGE: Native of Europe and Asia; scattered throughout the United States.

ILLINOIS DISTRIBUTION: Not common; scattered throughout Illinois.

The palea is nearly 2 mm shorter than the lemma, distinguishing this species rather tenuously from *B. arvensis*. In addition, the awns of *B. japonicus* tend to be flexuous.

14. Bromus inermis Leyss. Fl. Hal. 16. 1761. *Fig. 134.*

Bromus inermis var. *aristatus* Schur, Enum. Pl. Transsilv. 805. 1866.

Bromus inermis f. *aristatus* (Schur) Fern. Rhodora 35:316. 1933.

Perennial with culms to 1.2 m tall; sheaths glabrous; blades glabrous or nearly so, 5–15 mm broad; inflorescence erect, 10–27 cm long; spikelets 1.5–3.5 mm long, 5- to 10-flowered; glumes glabrous, acute, the first 1-nerved, 4–7 (–9) mm long, the second 3-nerved, 6–8 (–10) mm long; lemmas 3- to 7-nerved, 9–13 mm long, obtuse, glabrous or scabrous; awns absent, or rarely to 2 mm long; 2n = 56 (Wagnon, 1952).

COMMON NAME: Awnless Brome Grass; Hungarian Brome Grass.

HABITAT: Roadsides, fields, waste ground.

RANGE: Native of Europe; established in most of the United States.

ILLINOIS DISTRIBUTION: Common; probably in every county.

A few specimens which have awns up to 2 mm long have been observed in the Illinois collections. They may be

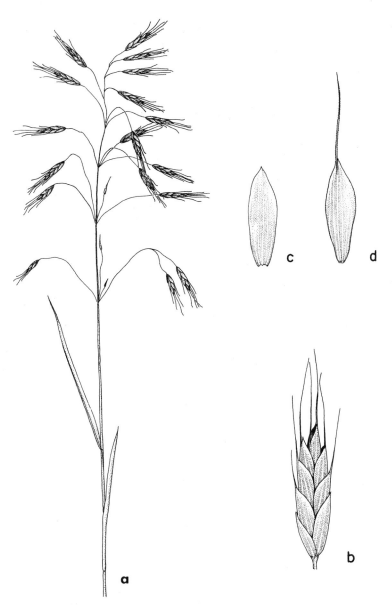

133. *Bromus japonicus* (Japonicus Chess). *a.* Inflorescence, X½. *b.* Spikelet, X2. *c.* Second glume, X4. *d.* Lemma, X4.

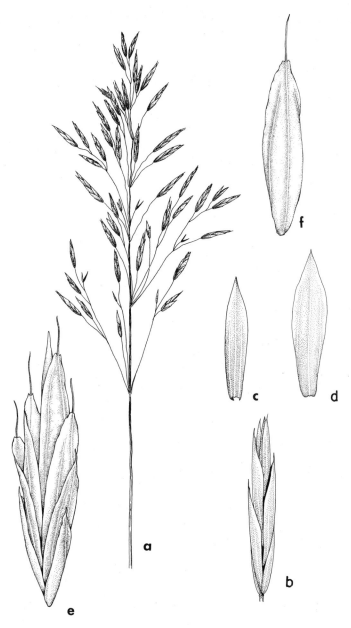

134. Bromus inermis (Awnless Brome Grass). *a.* Inflorescence, X½. *b.* Spikelet, X2. *c.* Second glume, X4. *d.* Lemma, X4. forma *aristata.* *e.* Spikelet, X4. *f.* Lemma, X5.

known as f. *aristatus*. The spikelets become purplish or bronze-colored at maturity. This species is variable in several respects: width of blades, length of inflorescence, length of spikelets, and number of veins per lemma.

The flowers are borne from May to July. This species appears to become more common each year.

15. Bromus erectus Huds. Fl. Angl. 39. 1762. *Fig. 135.*

Perennial with culms to 1 m tall; sheaths sparsely pilose to glabrous; blades sparsely pubescent on both surfaces, 2–3 mm broad, the lowermost usually plicate; inflorescence erect, narrow, 7–20 cm long; spikelets 1.5–3.0 cm long, 5- to 12-flowered; glumes glabrous or puberulent, subulate, the first 1-nerved, 6–9 mm long, the second 3-nerved, 9–11 mm long; lemmas 5- to 7-nerved, 11–13 mm long, glabrous or puberulent throughout; awns 4–7 mm long; 2n = 42 (Stählin, 1929).

COMMON NAME: Erect Brome Grass.

HABITAT: Waste ground.

RANGE: Native of Europe; infrequently established in the northeastern United States.

ILLINOIS DISTRIBUTION: Rare; only one collection seen from Illinois (St. Clair Co.: E. St. Louis, July 17, 1964, *R. H. Mohlenbrock 14226*).

The lone Illinois collection was made in mid-July.

16. Bromus purgans L. Sp. Pl. 76. 1753. *Fig. 136.*

Bromus altissimus Pursh, Fl. Am. Sept. 2:728. 1814, non Gilib. (1792).

Bromus ciliatus var. *purgans* (L.) Gray, Man. 600. 1848.

Bromus purgans latiglumis Shear, Bull. U.S.D.A. Div. Agrost. 23:40. 1900.

Bromus purgans incanus Shear, Bull. U.S.D.A. Div. Agrost. 23:41. 1900.

Bromus latiglumis (Shear) Hitchcock, Rhodora 8:211. 1906.

Bromus incanus (Shear) Hitchcock, Rhodora 8:212. 1906.

Bromus latiglumis f. *incanus* (Shear) Fern. Rhodora 35:316. 1933.

Cespitose perennial to 1.7 m tall; culms with 10–20 leaves; sheaths densely canescent-pilose to glabrous or nearly so; blades glabrous

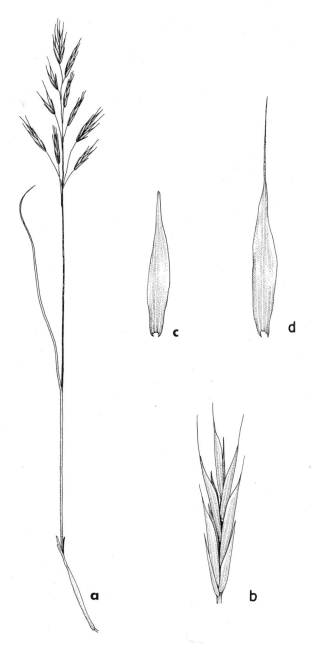

135. *Bromus erectus* (Erect Brome Grass). *a.* Inflorescence, X½. *b.* Spikelet, X2. *c.* Second glume, X4. *d.* Lemma, X4.

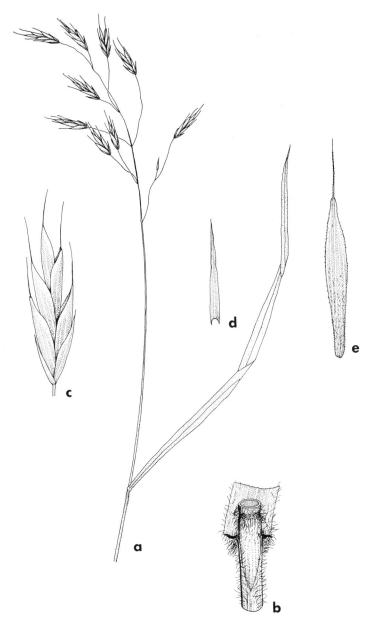

136. Bromus purgans (Brome Grass). *a.* Inflorescence, X½. *b.* Culm with sheath, X4. *c.* Spikelet, X2. *d.* First glume, X4. *e.* Lemma, X4.

or sparsely villous, 5–15 mm broad; inflorescence spreading or drooping, 15–30 cm long; spikelets 1.5–3.5 cm long, 3- to 8-flowered; glumes glabrous or pilose, the first 1-nerved, 5–8 mm long, the second 3-nerved, 6–10 mm long; lemmas strongly 5- to 7-nerved, 5–12 mm long, sericeous at base to nearly glabrous; awns 2–6 mm long.

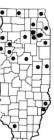

COMMON NAME: Brome Grass.

HABITAT: Usually moist, open woods.

RANGE: New Brunswick to Montana, south to New Mexico, Texas, Illinois, and North Carolina.

ILLINOIS DISTRIBUTION: Occasional in the northern half of the state, apparently rare in the southern half. This is the species which Fernald (1950), Gleason (1952), and others call *B. latiglumis*. Wagnon's studies (1952) indicate, however, that *B. purgans* is the correct name for this species.

Specimens with densely canescent-pilose sheaths have been known as *B. latiglumis* f. *incanus*, the type of which was collected by J. Wolf from Fulton County. There seems to be little reason for maintaining this form, since sheaths with all degrees of pubescence may be found.

This native species flowers from late June to mid-September.

17. Bromus pubescens Muhl. ex Willd. Enum. Pl. 120. 1809.
Fig. 137.

Bromus ciliatus var. *laeviglumis* Scribn. ex Shear, Bull. U.S.D.A. Div. Agrost. 23:32. 1900.
Bromus purgans f. *laevivaginatus* Wiegand, Rhodora 24:92. 1922.

Perennial with culms to 1.5 m tall; lower sheaths retrorsely pilose or glabrous, the upper glabrous or nearly so; blades hirtellous, sparsely villous, or glabrous, 5–15 mm broad; inflorescence spreading or drooping, 10–20 cm long; spikelets 2.0–3.5 cm long, 5- to 13-flowered; glumes pilose or glabrous, the first 1-nerved, 5–8 mm long, subulate, the second 3-nerved, 6–10 mm long, narrowly lanceolate; lemmas obscurely 5- to 7-nerved, 8–12 mm long, pubescent throughout; awns 2–8 mm long; 2n =14, 28 (Elliott, 1949, as *B. purgans*).

137. Bromus pubescens (Canada Brome Grass). *a.* Inflorescence, X½. *b.* Spikelet, X2. *c.* Second glume, X4. *d.* Lemma, X4.

COMMON NAME: Canada Brome Grass.

HABITAT: Moist, open woods.

RANGE: Quebec to Alberta, south to Texas and Florida.

ILLINOIS DISTRIBUTION: Common throughout the state.

This species, through misunderstanding of the type (according to Wagnon [1952]), has been referred to as *B. purgans* by many authors.

Most specimens have the lower sheaths retrorsely pilose; a collection from Cook County has the sheaths entirely glabrous.

Bromus pubescens flowers from June to August.

18. Bromus ciliatus L. Sp. Pl. 76. 1753. *Fig. 138.*

Bromus canadensis Michx. Fl. Bor. Am. 1:65. 1803.

Bromus ciliatus var. *intonsus* Fern. Rhodora 32:70. 1930.

Perennial with culms to 1.3 m tall; sheaths glabrous or scabrous, or the middle and upper retrorsely pilose; blades hirtellous, sparsely pilose, glabrous or scabrous on both surfaces, 5–15 mm broad; inflorescence drooping, 10–25 cm long; spikelets 1–3 cm long, 3- to 10-flowered; glumes conduplicate, the first 1-nerved, 5–7 mm long, the second 3-nerved, 7–10 mm long; lemmas 5- to 7-nerved, 10–12 mm long, conduplicate or involute, pubescent on the margins in the lower two-thirds; awns 3–5 mm long; $2n = 14$, 28 (Elliott, 1949).

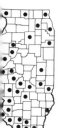

COMMON NAME: Canada Brome Grass.

HABITAT: Open woodlands.

RANGE: Labrador to British Columbia, south to California, Texas, and New Jersey.

ILLINOIS DISTRIBUTION: Common throughout Illinois.

The majority of the Illinois specimens have the middle and upper sheaths retrorsely pilose; these have been known as var. *intonsus*. Other variations may be found in the width of the leaves, the length of the inflorescence, and number of flowers per spikelet. The spikelets occasionally become purplish at maturity.

This species flowers from late June to late September.

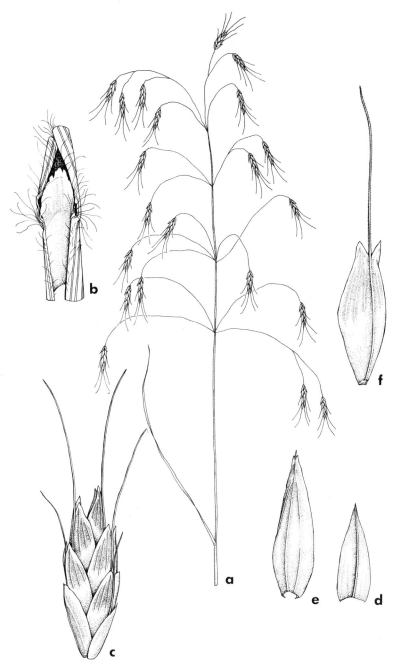

138. *Bromus ciliatus* (Canada Brome Grass). *a.* Inflorescence, X½. *b.* Sheath, with ligule, X5. *c.* Spikelet, X4. *d.* First glume, X5. *e.* Second glume, X5. *f.* Lemma, X5.

2. *Vulpia* K. C. GMEL.

Annuals; blades narrow, often involute; inflorescence paniculate, usually ascending; spikelets 3- to several-flowered, disarticulating above the glumes; glumes 2, unequal, shorter than the spikelets; lemmas convex, obscurely 5-nerved, usually awned; stamen 1.

Vulpia often is treated as a section of *Festuca* and this, indeed, may be its proper disposition. It differs from *Festuca* in its annual habit and its single stamen.

KEY TO THE SPECIES OF *Vulpia* IN ILLINOIS

1. Awns of lemma, if present, up to 5.5 mm long; first glume one-half to nearly equalling the second glume_____1. *V. octoflora*
1. Some or all of the awns of the lemma 1 cm long or longer; first glume about one-fourth as long as the second glume__2. *V. myuros*

1. *Vulpia octoflora* (Walt.) Rydb. Bull. Torrey Club 36:538. 1909.

Festuca octoflora Walt. Fl. Carol. 81. 1788.

Slender, erect annual to 50 cm tall; sheaths glabrous; blades involute, glabrous, about 1 mm broad; inflorescence narrow, 3–12 cm long; spikelets 5–12 mm long, 5- to 13-flowered; glumes narrowly lanceolate, the first 1-nerved, 1.5–4.5 mm long, the second 3-nerved, 3.0–5.5 mm long; lemmas involute, lanceolate, glabrous or scabrous, 2.5–5.2 mm long; awns 1.0–5.5 mm long, sometimes absent.

Three fairly distinct varieties of *V. octoflora* occur in Illinois. A discussion of these varieties has been presented by Fernald (1945).

1. Awns 3.5–5.5 mm long; inflorescence more or less appearing racemose_____1a. *V. octoflora* var. *octoflora*
1. Awns 1–3 mm long, sometimes absent; inflorescence loosely or densely spicate.
    2. Awns 1–3 mm long; lower glume 2.5–4.0 mm long; inflorescence loosely spicate_____1b. *V. octoflora* var. *tenella*
    2. Awns absent or up to 2 mm long; lower glume 1.5–3.0 mm long; inflorescence densely spicate_____1c. *V. octoflora* var. *glauca*

1a. *Vulpia octoflora* Walt. var. *octoflora* *Fig. 139.*

Inflorescence appearing racemose; lower glume 3.5–4.5 mm long; awns 3.5–5.5 mm long.

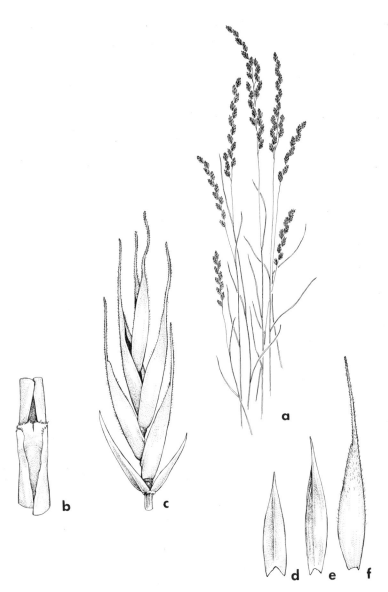

139. *Vulpia octoflora* var. *octoflora* (Six-weeks Fescue). *a.* Habit, X½. *b.* Sheath, with ligule, X6. *c.* Spikelet, X10. *d.* First glume, X10. *e.* Second glume, X10. *f.* Lemma, X10.

COMMON NAME: Six-weeks Fescue.

HABITAT: Dry soil.

RANGE: New Jersey to Oklahoma, south to Texas and Florida.

ILLINOIS DISTRIBUTION: Scattered throughout the state. This variety and var. *tenella* are about equally abundant in Illinois.

All three varieties of *V. octoflora* in Illinois flower from May to early July.

The length of the awn (3.5–5.5 mm) in this variety serves best to distinguish it from the other two varieties. Variety *octoflora* usually has a more open inflorescence.

1b. Vulpia octoflora Walt. var. **tenella** (Willd.) Fern. Rhodora 47:107. 1945. *Fig. 140.*

Festuca tenella Willd. Sp. Pl. 1:419. 1797.

Vulpia tenella (Willd.) Heynh. Nom. 1:854. 1840.

Festuca octoflora var. *tenella* (Willd.) Fern. Rhodora 34:209. 1932.

Inflorescence loosely spicate; lower glume 2.5–4.0 mm long; awns 1–3 mm long.

HABITAT: Dry soil.

RANGE: Quebec to British Columbia, south to California, Texas, and Georgia.

ILLINOIS DISTRIBUTION: Rather common; scattered throughout the state.

Considerable difficulty may be encountered in distinguishing this variety from var. *glauca*. In general, var. *tenella* is slightly larger in all respects.

1c. Vulpia octoflora Walt. var. **glauca** (Nutt.) Fern. Rhodora 47:107. 1945. *Fig. 141.*

Festuca tenella var. *glauca* Nutt. Amer. Phil. Soc. Trans. 5:147. 1837.

Festuca octoflora var. *glauca* (Nutt.) Fern. Rhodora 34:209. 1932.

Inflorescence densely spicate; lower glume 1.5–3.0 mm long; awns absent or up to 2 mm long.

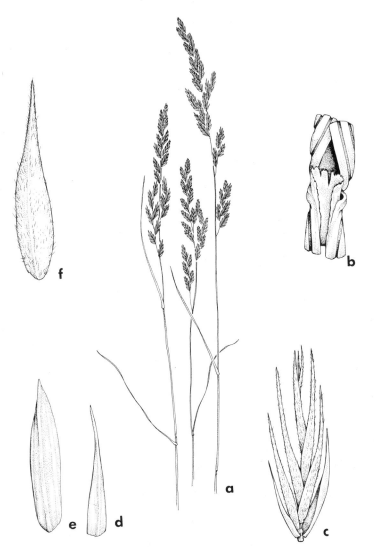

140. Vulpia octoflora var. *tenella* (Six-weeks Fescue). *a.* Habit, X½. *b.* Sheath, with ligule, X6. *c.* Spikelet, X10. *d.* First glume, X10. *e.* Second glume, X10. *f.* Lemma, 10.

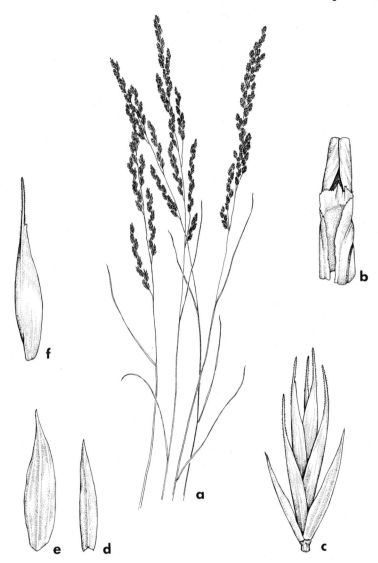

141. Vulpia octoflora var. *glauca* (Six-weeks Fescue). *a.* Habit, X½. *b.* Sheath, with ligule, X6. *c.* Spikelet, X10. *d.* First glume, X10. *e.* Second glume, X10. *f.* Lemma, X10.

HABITAT: Dry soil.

RANGE: Indiana to Wyoming, south to New Mexico and Florida. This is the most southern of the three varieties.

ILLINOIS DISTRIBUTION: The least common of the varieties of *V. octoflora* in Illinois.

In their extreme conditions, the three varieties are rather easily distinguishable.

2. **Vulpia myuros** (L.) K. Gmel. Fl. Badens. 1:8. 1805. *Fig. 142.*

Festuca myuros L. Sp. Pl. 74. 1753.

Erect annual to 65 cm tall; sheaths glabrous; blades involute, glabrous, about 1 mm broad; inflorescence to 20 cm long, the spikelets ascending; spikelets 3- to 5-flowered; glumes linear to narrowly lanceolate, the first 1-nerved, 1.0–1.5 mm long, about ¼ as long as the second, the second 3-nerved, 4–5 mm long; lemmas involute, linear-lanceolate, scabrous on the back, 4.5–7.0 mm long; awns 8–16 mm long.

COMMON NAME: Foxtail Fescue.

HABITAT: Dry fields (in Illinois).

RANGE: Native of Europe; occasionally collected as an adventive in the United States.

ILLINOIS DISTRIBUTION: Known only from Johnson County (Wildcat Bluff, June 22, 1969, *J. A. White 1274, 1275*) and Massac County (west shore of Hohman Lake, June 6, 1970, *John Schwegman s.n.*).

The very long awns and the extremely short first glume readily distinguish this more robust *Vulpia* from *V. octoflora.*

3. *Festuca* L. — Fescue

Perennials; blades plicate, flat, or involute; inflorescence paniculate, ascending, spreading, or nodding; spikelets 2- to 13-flowered, disarticulating above the glumes; glumes 2, unequal, shorter than the spikelets; lemmas convex or involute, obscurely nerved or nerveless, with or without an awn; palea shorter than to equaling the lemmas; stamens 3.

142. Vulpia myuros (Foxtail Fescue). *a.* Inflorescences, X½. *b.* Sheath, with ligule, X6. *c.* Spikelet, X10. *d.* Glumes, X17½.

Festuca is distinguished from *Bromus* by the lack of a bidentate apex of the lemmas.

Both native and introduced species of fescue occur in Illinois. Most of the introduced species are escaped from lawns. Both *F. ovina* and *F. rubra* are good grazing grasses in the western United States.

Two rather old treatments of *Festuca* are by Piper (1906) and St. Yves (1926).

KEY TO THE SPECIES OF Festuca IN ILLINOIS

1. Leaf blades involute or plicate, 0.4–1.2 mm broad.
    2. First glume 1.2 mm long; second glume 1.8–3.0 mm long; lemmas 2.5–3.5 mm long; awns absent, or to 0.5 mm long_____ _____1. *F. capillata*
    2. First glume 2.5–4.5 mm long; second glume 3.5–5.5 mm long; lemmas 4–7 mm long; awns 1.0–3.5 mm long.
        3. Lowest sheaths whitish, not becoming fibrous; lemmas essentially nerveless_____2. *F. ovina*
        3. Lowest sheaths brown or reddish, becoming fibrous; lemmas 3- to 5-nerved_____3. *F. rubra*
1. Leaf blades flat, 3–11 mm broad.
    4. Lemmas 5.5–10.0 mm long.
        5. Inflorescence 6- to 11-flowered; lemmas 5.5–8.0 mm long__ _____4. *F. pratensis*
        5. Inflorescence 4- to 5-flowered; lemmas 7–10 mm long_____ _____5. *F. arundinacea*
    4. Lemmas 3.3–5.2 mm long.
        6. Inflorescence spreading at maturity; spikelets to 4 mm broad; lemmas acute or subacute_____6. *F. obtusa*
        6. Inflorescence ascending at maturity; spikelets about 5 mm broad; lemmas obtuse_____7. *F. paradoxa*

1. **Festuca capillata** Lam. Pl. Franc. 3:597. 1778. *Fig. 143.*

Festuca ovina var. *capillata* (Lam.) Alefeld, Landw. Fl. 354. 1866.

Densely tufted perennial to 50 cm tall; sheaths glabrous, whitish, not becoming fibrous; blades capillary, involute, less than 1 mm in diameter; inflorescence linear to oblong, 1–7 cm long; spikelets 4–6 mm long, 4- to 6-flowered; glumes glabrous, the first 1–2 mm long, the second 1.8–3.0 mm long; lemmas involute, coriaceous, 2.5–3.5 mm long; awns absent, or up to 0.5 mm long.

143. Festuca capillata (Slender Fescue). *a.* Habit, X½. *b.* Sheath, with ligule, X6. *c.* Spikelet, X10. *d.* First glume, X10. *e.* Second glume, X10.

COMMON NAME: Slender Fescue.

HABITAT: Waste ground.

RANGE: Native of Europe and probably Newfoundland; occasionally escaped and established in northeastern North America.

ILLINOIS DISTRIBUTION: Not common; scattered throughout the state in the metropolitan areas.

This species flowers from the last of May to mid-July. Gleason (1952) considers this taxon to be merely a variety of *F. ovina*, but the very short awns in *F. capillata* are very distinctive. This species, as well as the next, is discussed thoroughly by Fernald (1935). The short glumes separate *F. capillata* from other fescues.

2. **Festuca ovina** L. var. **duriuscula** (L.) Koch, Syn. Fl. Germ. Helv. 812. 1837. *Fig. 144.*

Festuca duriuscula L. Sp. Pl. 74. 1753.

Densely tufted perennial to 50 cm tall; sheaths glabrous, whitish, not becoming fibrous at maturity; blades involute, glabrous or scabrous, 0.7–1.2 mm in diameter; inflorescence narrow, 5–8 cm long; spikelets 5–10 mm long, 4- to 9-flowered; glumes attenuate, glabrous, the first 2.5–4.0 mm long, the second 3.5–5.0 mm long; lemmas involute, coriaceous, nerveless, 4–6 mm long; awns 1.3–3.0 mm long.

COMMON NAME: Sheep Fescue.

HABITAT: Waste ground.

RANGE: Native of Europe; established in the northeastern United States.

ILLINOIS DISTRIBUTION: Not common, but scattered throughout the state.

Typical var. *ovina*, which is introduced in more northern regions of the United States, has more slender leaves and smaller panicles and spikelets. The plants generally have a gray-green color.

3. **Festuca rubra** L. Sp. Pl. 74. 1753. *Fig. 145.*

Loosely tufted perennial, decumbent at the base, with culms to 1 m tall; lowest sheaths brown or reddish, becoming fibrous at maturity, glabrous or puberulent; blades plicate or involute, to 1 mm broad; inflorescence narrow, ascending, 3–20 cm long; spikelets 7–12 mm long, 4- to 7-flowered; glumes glabrous, the first subu-

144. *Festuca ovina* var. *duriuscula* (Sheep Fescue). *a.* Inflorescences, X¾. *b.* Sheath, with ligule, X5. *c.* Spikelet, X15. *d.* First glume, X15. *e.* Second glume, X15. *f.* Lemma and palea, X15.

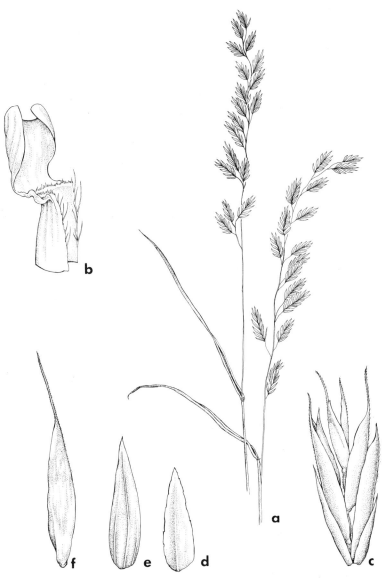

145. Festuca rubra (Red Fescue). *a.* Inflorescences, X¾. *b.* Sheath, with ligule, X5. *c.* Spikelet, X5. *d.* First glume, X7½. *e.* Second glume, X7½. *f.* Lemma, X7½.

late, 2.5–4.5 mm long, the second narrowly lanceolate, glabrous or scabrous, 3- to 5-nerved, 5–7 mm long; awns 1.0–3.5 mm long.

COMMON NAME: Red Fescue.

HABITAT: Waste ground or cultivated areas.

RANGE: Greenland to Alaska, south to California, Texas, and Georgia; Mexico; Europe; Asia; Africa.

ILLINOIS DISTRIBUTION: Infrequent throughout the state. First collected in Illinois by Virginius Chase in 1948 from Peoria County. The reddish (or brownish) lower sheaths, which account for the specific epithet, distinguish *F. rubra* from either *F. ovina* or *F. capillata*.

4. **Festuca pratensis** Huds. Fl. Angl. 37. 1762. *Fig. 146.*

Festuca elatior sensu auct., non L. 1753, *sensu strictu.*

Festuca elatior L. var. *pratensis* (Huds.) Gray, Man., ed. 4. 634. 1867.

Festuca elatior f. *aristata* Holmb. Skand. Fl. 1927.

Loosely tufted perennial to 1.2 m tall; sheaths glabrous; blades flat, glabrous or scabrous above, 4–8 mm broad; inflorescence erect or nodding, 5–25 cm long; spikelets 8–12 mm long, 6- to 11-flowered; glumes glabrous, the first subulate, 2.5–4.5 mm long, the second lanceolate, 3.5–7.0 mm long; lemmas lanceolate, glabrous, obscurely nerved, 5.5–8.0 mm long; awns absent, or rarely to 1 mm long; 2n = 14 (Myers & Hill, 1947).

COMMON NAME: Meadow Fescue.

HABITAT: Waste ground.

RANGE: Native of Europe; introduced throughout the United States.

ILLINOIS DISTRIBUTION: Common; in every county. Although this species has been universally known as *F. elatior,* Terrell (1967) presents evidence that the correct binomial should be *F. pratensis.*

This plant is abundant along railroads where it flowers from late May to mid-August. It is planted frequently by the highway department for a quick ground-cover along new roads.

The first Illinois collections were apparently made during the 1860's.

Specimens with awns about 1 mm long are known.

146. *Festuca pratensis* (Meadow Fescue). *a.* Habit, X½. *b.* Sheath, with ligule, X5. *c.* Spikelet, X5. *d.* First glume, X10. *e.* Second glume, X10. *f.* Lemma, X10.

5. **Festuca arundinacea** Schreb. Spic. Fl. Lips. 57. 1771. *Fig. 147.*

Festuca elatior L. Sp. Pl. 75. 1753, *sensu strictu; nomen ambiguum rejiciendum.*

Festuca elatior L. var. *arundinacea* (Schreb.) Wimm. Fl. Schles. 59, ed. 3. 1857.

Loosely tufted perennial to 1.5 m tall; sheaths glabrous; blades glabrous or scabrous, flat, 4–8 mm broad; inflorescence spreading to ascending, 15–30 cm long; spikelets 6–11 mm long, 4- to 5-flowered; glumes glabrous, the first subulate, 3–5 mm long, the second lanceolate, 4.5–8.0 mm long; lemmas lanceolate, glabrous, obscurely nerved, 7–10 mm long; awns absent; 2n = 42 (Myers & Hill, 1947), 28 (Stählin, 1929).

COMMON NAME: Large Fescue.

HABITAT: Waste ground.

RANGE: Native of Europe; sometimes established in the northern United States.

ILLINOIS DISTRIBUTION: Rare; known from Jackson, St. Clair, Hardin, Peoria, Union, and Tazewell counties.

6. **Festuca obtusa** Biehler, Pl. Nov. Herb. Spreng. Cent. 11. 1807. *Fig. 148.*

Panicum divaricatum Michx. Fl. Bor. Am. 1:50. 1803, non L. (1753).

Tufted perennial to 1 m tall; sheaths glabrous or puberulent; blades flat, glabrous, scabrous on the veins, 3–11 mm broad; inflorescence spreading at maturity, 12–25 cm long; spikelets 4–8 mm long, up to 4 mm broad, 2- to 5-flowered; glumes glabrous with hyaline margins, the first subulate, 2–4 mm long, the second ovate, 2.5–4.5 mm long; lemmas glabrous, acute, obscurely nerved, 3.0–4.5 mm long; awns none.

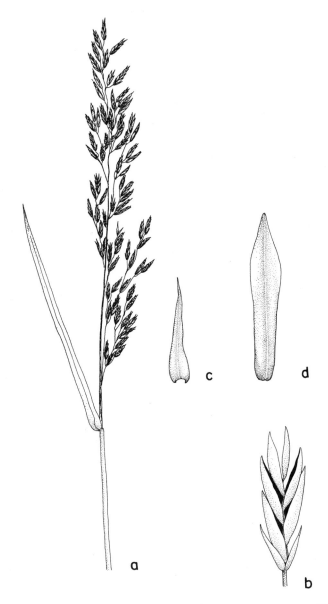

147. Festuca arundinacea (Large Fescue). *a.* Inflorescence, X½. *b.* Spikelet, X4. *c.* Glume, X8. *d.* Lemma, X8.

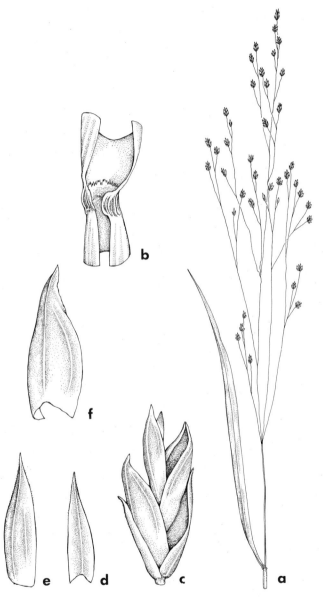

148. Festuca obtusa (Nodding Fescue). *a.* Inflorescence, X½. *b.* Sheath, with ligule, X5. *c.* Spikelet, X12½. *d.* First glume, X15. *e.* Second glume, X15. *f.* Lemma, X15.

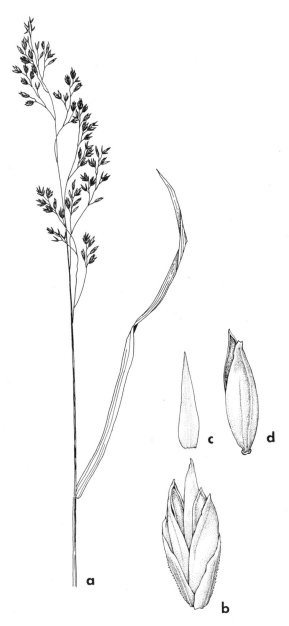

149. *Festuca paradoxa* (Fescue). *a.* Inflorescence, X½. *b.* Spikelet, X4.
c. Second glume, X8. *d.* Lemma, X8.

COMMON NAME: Nodding Fescue.

HABITAT: Moist woodlands.

RANGE: Nova Scotia to Manitoba, south to Texas and Florida.

ILLINOIS DISTRIBUTION: Occasional; scattered throughout the state and very possibly in every county.

The Nodding Fescue flowers in May, June, and July. It is not always easily distinguishable from *F. paradoxa,* but the acute lemmas of *F. obtusa* are most reliable. Considerable variation exists in the width of the blades.

Until 1918, Illinois botanists referred to this species as *F. nutans,* but this clearly is not the same species as *F. nutans* Moench.

7. Festuca paradoxa Desv. Opusc. 105. 1831. *Fig. 149.*

Festuca nutans Bieler, Pl. Nov. Herb. Spreng. Cent. 10. 1807, non Moench (1794).

Festuca shortii Kunth ex Wood, Class-book 794. 1861.

Tufted perennial to 1.2 m tall; sheaths glabrous; blades flat, glabrous or scabrous, 4–8 mm broad; inflorescence ascending at maturity, 12–20 cm long; spikelets 5–8 mm long, about 5 mm broad, 3- to 6-flowered; glumes glabrous, the first subulate, 2.5–4.0 mm long, the second elliptic, 3–5 mm long; lemmas glabrous, obtuse, obscurely nerved, 3.5–5.0 mm long; awns none.

COMMON NAME: Fescue.

HABITAT: Dry or moist woodlands.

RANGE: Maryland to Minnesota, south to Texas and Georgia.

ILLINOIS DISTRIBUTION: Occasional in the southern three-fourths of the state; apparently absent from the northern one-fourth.

The height of the plant and the length of the glumes and lemmas are slightly larger than *F. obtusa;* in addition, the lemmas of *F. paradoxa* are obtuse.

This species flowers from mid-May to July. It seems to be more plentiful in dry, rocky woodlands than in more moist situations.

4. Lolium L. — Rye Grass

Annuals or perennials; blades flat; inflorescence spicate, erect; spikelets several-flowered, solitary and placed edgewise at each joint of the rachis, fitting into the flexuous rachis, disarticulating above the glumes; glume 1 (2 in the terminal spikelet), strongly nerved; lemmas rounded on the back, several-nerved, awned or awnless.

The genus is recognized easily by the spikelets which are placed edgewise at each joint of the rachis.

Traditionally *Lolium* has been placed in tribe Hordeae. Recent studies have shown that *Lolium* and *Festuca* apparently are closely related. Indeed, hybrids are known between the two.

A revision of *Lolium* has been prepared by Terrell (1968).

KEY TO THE SPECIES OF Lolium IN ILLINOIS

1. Glume as long as or longer than the spikelet, 15–20 mm long_____
_____1. *L. temulentum*
1. Glume shorter than the spikelet, 4–12 mm long.
    2. Lemmas (or at least the uppermost) awned; spikelets 10- to 20-flowered; annual_____2. *L. multiflorum*
    2. Lemmas awnless; spikelets 6- to 10-flowered; perennial_____
_____3. *L. perenne*

1. Lolium temulentum L. Sp. Pl. 83. 1753. *Fig. 150.*

Annual with culms to 75 cm tall; blades 3–6 mm broad, scaberulous; spike 15–25 cm long; spikelets 5- to 7-flowered; glume as long as or longer than the spikelet, 15–20 mm long, acuminate, 5- to 7-nerved; lemmas 5–8 mm long, obtuse, with an awn 5–12 mm long; 2n = 14 (Jenkin & Thomas, 1938).

COMMON NAME: Darnel.

HABITAT: Waste ground, fields.

RANGE: Native of Europe; introduced throughout the United States.

ILLINOIS DISTRIBUTION: Not common.

This species flowers during June and July in Illinois. It is reputed to contain a poisonous narcotic due to the presence of a fungus. Hall's collection from Menard County in 1861 is the first from Illinois.

The very long glumes differentiate this species from all other species of *Lolium* in Illinois. Variation exists in the length of the awns of the lemmas.

2. Lolium multiflorum Lam. Fl. Franc. 3:621. 1778. *Fig. 151.*

Lolium italicum A. Br. Flora 17:241. 1834.

Lolium perenne var. *multiflorum* (Lam.) Parnell, Grasses Brit. 302. 1845.

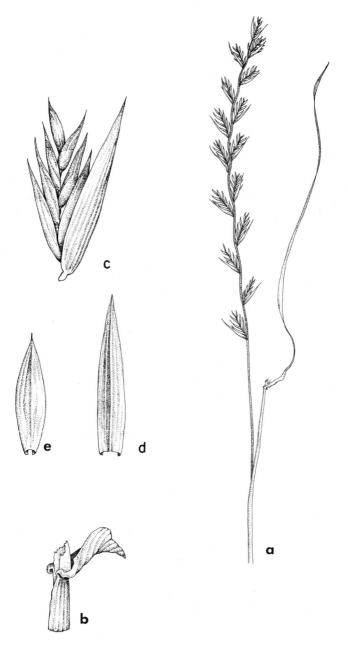

150. *Lolium temulentum* (Darnel). *a.* Inflorescence, X½. *b.* Sheath, with ligule, X5. *c.* Spikelet, X3½. *d.* Glume, X2¾. *e.* Lemma, X2¾.

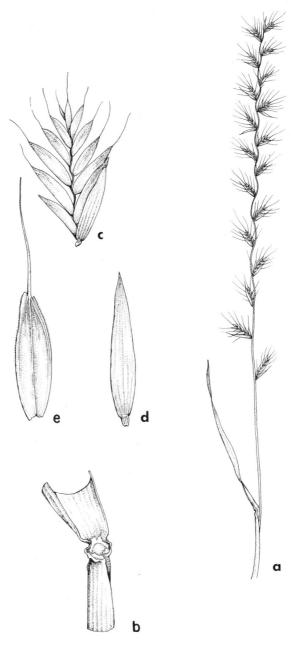

151. *Lolium multiflorum* (Italian Rye Grass). *a.* Inflorescence, X½. *b.* Sheath, with ligule, X5. *c.* Spikelet, X2½. *d.* Glume, 4. *e.* Lemma, X4.

Lolium temulentum var. *multiflorum* (Lam.) Kuntze, Rev. Gen. Pl. 2:779. 1891.

Annual with culms to 75 cm tall; blades 2–4 mm broad, scaberulous; spike 15–25 cm long; spikelets 10- to 20-flowered; glume shorter than the spikelet, 4–12 mm long, 5- to 7-nerved; lemmas progressively smaller from base to summit, the largest 5.5–8.0 mm long, at least the upper with awns to 8 mm long; 2n = 14 (Peto, 1933).

COMMON NAME: Italian Rye Grass.

HABITAT: Waste ground, fields.

RANGE: Native of Europe; introduced throughout the United States.

ILLINOIS DISTRIBUTION: Occasional; throughout the state. This species flowers from late May to early September. The shortened glumes relate this species to *L. perenne*, but *L. multiflorum* has more flowers per spikelet and has some of the lemmas awned.

Apparently the first collection made of this species in Illinois was by Clokey from Macon County in 1898.

3. **Lolium perenne** L. Sp. Pl. 83. 1753. *Fig. 152.*

Perennial with culms to 60 cm tall; blades 2–4 mm broad, glabrous or nearly so; spike 10–20 cm long; spikelets 6- to 10-flowered; glume shorter than the spikelet, 6–12 mm long, 5- to 7-nerved; lemmas progressively smaller from base to summit, the largest 5–7 mm long, awnless; 2n = 14 (Thomas, 1936).

COMMON NAME: English Rye Grass.

HABITAT: Waste ground, fields, lawns.

RANGE: Native of Europe; introduced throughout the United States.

ILLINOIS DISTRIBUTION: Common; probably in every county.

This species is commonly used in the seeding of new lawns. It escapes freely. It flowers from June to September. It is the most common *Lolium* in Illinois. The complete lack of awns of the lemma is unique among the Illinois species of *Lolium*. The length of the glume ranges from 6–12 mm.

Hybrids between *L. multiflorum* and *L. perenne* may occur in Illinois.

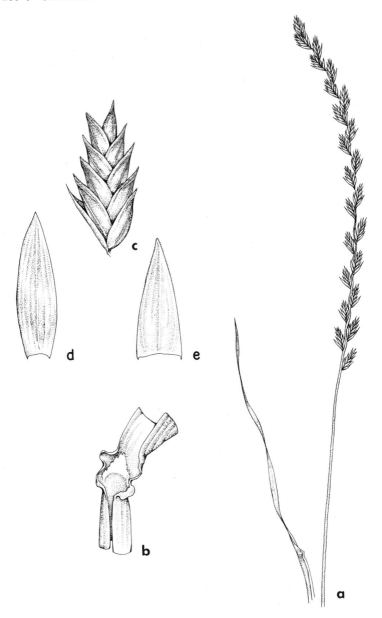

152. Lolium perenne (English Rye Grass). *a.* Inflorescence, X½. *b.* Sheath, with ligule, X5. *c.* Spikelet, X3. *d.* Glume, X5. *e.* Lemma, X5.

5. *Puccinellia* PARL.

Glabrous annuals or perennials; sheaths open; blades involute or flat; inflorescence paniculate; spikelets several-flowered, disarticulating above the glumes; glumes 2, slightly unequal, shorter than the spikelets; lemmas rounded on the back, obscurely or distinctly 5-nerved, awnless; lodicules free from each other; stamens 3; style absent.

Most species of this genus grow in saline situations. One of our representatives (*P. pallida*) until recently was known as a *Glyceria*. *Glyceria* differs by its closed sheaths.

KEY TO THE SPECIES OF Puccinellia IN ILLINOIS

1. Lemmas obscurely nerved; plants adventive in Illinois along railroad tracks or highways_____1. *P. distans*
1. Lemmas conspicuously nerved; plants native in Illinois in swamps in extreme southern Illinois_____2. *P. pallida*

1. Puccinellia distans (L.) Parl. Fl. Ital. 367. 1848. *Fig. 153.*

Poa distans L. Mant. Pl. 1:32. 1767.
Glyceria distans (L.) Wahl. Fl. Upsal. 36. 1820.
Panicularia distans (L.) Kuntze, Rev. Gen. Pl. 2:782. 1891.
Tufted perennial, decumbent at the base, with the culms to 75 cm tall; blades flat to involute, up to 4 mm broad; inflorescence paniculate, some of the branches reflexed, 5–20 cm long; spikelets 4–6 mm long, 4- to 7-flowered, green or purplish; glumes glabrous, ovate, the first 0.8–1.2 mm long, the second 1.2–2.0 mm long; lemmas narrowly ovate, more or less obtuse, ciliate along the margins, obscurely nerved, 1.5–2.5 mm long; 2n = 14 (Avdulov, 1931).

COMMON NAME: Alkali Grass.
HABITAT: Waste ground.
RANGE: Native of Europe; adventive in North America along the Atlantic Coast and near the Great Lakes.
ILLINOIS DISTRIBUTION: Rare; known only from Cook County.
The first collection of this species in Illinois was in 1957.

153. *Puccinellia distans* (Alkali Grass.) *a.* Habit, X¼. *b.* Sheath, with ligule, X5. *c.* Spikelet, front view, X10. *d.* Spikelet, back view, X10.

2. **Puccinellia pallida** (Torr.) Clausen, Rhodora 54:44. 1952.
   Fig. 154.

Windsoria pallida Torr. Cat. Pl. N. Y. 91. 1819.
Glyceria pallida (Torr.) Trin. Acad. St. Petersb. Mem. VI. Sci.
Nat. 2:57. 1836.
Panicularia pallida (Torr.) Kuntze, Rev. Gen. Pl. 2:783. 1891.
Torreyochloa pallida (Torr.) Church, Amer. Jour. Bot. 36:164.
1949.

Semi-aquatic perennial with decumbent bases and culms to 1 mm
tall; leaves soft, 3–8 mm broad; inflorescence paniculate, 5–25 cm
long; spikelets 4–7 mm long, 4- to 8-flowered, pale green; glumes
glabrous, ovate, the first 1–2 mm long, the second 1.2–2.5 mm
long; lemmas narrowly ovate, obtuse, puberulent near the apex,
5- to 7-nerved, 2.0–3.5 mm long; 2n = 14 (Church, 1949).

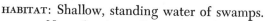

HABITAT: Shallow, standing water of swamps.

RANGE: Nova Scotia to Ontario, south to Missouri and
Virginia; Indiana.

ILLINOIS DISTRIBUTION: Rare; known only from Union
County.

Clausen (1952) has suggested that this species belongs
in the genus *Puccinellia,* rather than in *Glyceria* where
it has been placed since 1836. Clausen's reasoning is followed in this treatment.

This species grossly resembles the genus *Glyceria* rather
than *Puccinellia,* but is obviously most nearly related to other
Puccinellia species on the basis of diagnostic characters. There is
a strong resemblance superficially between this species and *Glyceria arkansana* and, indeed, the two grow together in the LaRue
Swamp of Union County.

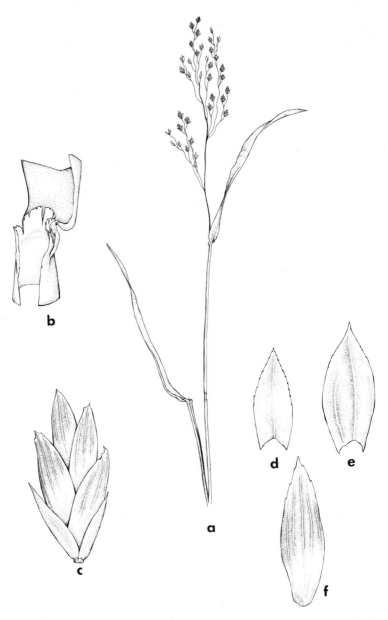

154. *Puccinellia pallida.* *a*. Inflorescence, X½. *b*. Sheath, with ligule, X5.
c. Spikelet, X12½. *d*. First glume, X17½. *e*. Second glume, X17½. *f*.
Lemma, X17½.

6. *Poa* L. – Bluegrass

Annuals or tufted or rhizomatous perennials; blades flat, mostly boat-shaped at the tips; inflorescence paniculate; spikelets several-flowered, disarticulating above the glumes; glumes 2, more or less unequal, shorter than the spikelets; lemmas distinctly keeled and nerved, awnless, usually with a tuft of cobwebby hairs at the base (except *P. annua* and *P. autumnalis*).

The tuft of cobwebby hairs at the base of the lemmas (except in two species) sets this genus apart from all others in the tribe Festuceae.

Both native and introduced species of *Poa* occur in Illinois. Four sections are represented: Annuae, the plants annual (*P. annua, P. chapmaniana*); Pratenses, the plants perennial with rhizomes (*P. compressa, P. arachnifera, P. angustifolia, P. pratensis*); Alpinae, the plants perennial, tufted, without a webbed lemma (*P. autumnalis*); and Palustres, the plants perennial, tufted, with a webbed lemma (the remainder of the species).

Some species, particularly *P. pratensis* (Kentucky Bluegrass), are choice lawn grasses.

Although *P. cuspidata* has been attributed to Illinois by several workers, it is excluded by Hitchcock (1950), and no Illinois material has been seen.

Species determination mainly rests with the pubescence or lack of it on the lemmas. Two species lack the tuft of cobwebby hairs at the base of the lemma. In the fifteen species of *Poa* in Illinois, four basic patterns of pubescence of the nerves and keel of the lemmas are represented. In seven species, all the nerves as well as the keel are pubescent; in four species, the distant marginal nerves and the keel are pubescent, but the obscure, intermediate nerves are glabrous; in three species, only the keel is pubescent; in one species, neither the nerves nor the keel is pubescent.

KEY TO THE SPECIES OF Poa IN ILLINOIS

1. Plants dioecious; pistillate spikelets woolly; staminate spikelets glabrous or nearly so_____4. *P. arachnifera*
1. Plants monoecious or spikelets perfect, variously pubescent or glabrous.
    2. Lemmas without a tuft of cobwebby hairs at the base.
        3. Tufted annual to about 30 cm tall, sometimes rooting at the lower nodes; lemmas elliptic to ovate_____1. *P. annua*

3. Tufted perennial to 75 cm tall, not rooting at the lower nodes;
   lemmas oblong_____3. *P. autumnalis*
2. Lemmas with a tuft of cobwebby hairs at the base.
   4. Nerves and keel of the lemma glabrous (except for the cob-
      webby tuft)_____8. *P. languida*
   4. Keel and sometimes the nerves of the lemma pubescent.
      5. Keel of the lemma pubescent, the nerves glabrous.
         6. Culms beneath the panicle and the sheaths scabrous;
            lemmas sharply nerved; ligule of upper leaves 4–8 mm
            long_____9. *P. trivialis*
         6. Culms beneath the panicle and the sheaths usually gla-
            brous; lemmas obscurely nerved; ligule of upper leaves
            about 1 mm long_____10. *P. alsodes*
      5. Keel and at least some of the nerves of the lemma pubes-
         cent.
         7. Marginal nerves of the lemma pubescent, the interme-
            diate nerves glabrous.
            8. Plants with rhizomes; lemmas with 5 prominent
               nerves.
               9. Basal leaves flat, at least as broad as the culm;
                  culm compressed at base, 2–3 mm thick at base;
                  glumes broadly lanceolate, straight_____
                  _____5. *P. pratensis*
               9. Basal leaves involute or filiform, narrower than
                  the culm; culm terete at base, 1–2 mm thick at
                  base; glumes narrowly lanceolate, arching_____
                  _____6. *P. angustifolia*
            8. Plants without rhizomes; lemmas with 3 prominent
               nerves and 2 obscure nerves.
               10. Culms very weak, solitary or in small tufts;
                   sheaths scabrous; lowest branches of the panicle
                   mostly paired_____11. *P. paludigena*
               10. Culms more firm, usually densely tufted; sheaths
                   usually glabrous; lowest branches of the panicle
                   in clusters of 3–5.
                   11. Ligule 0.5–1.0 mm long; anthers 1.2–1.6
                       mm long_____12. *P. nemoralis*
                   11. Ligule 2–5 mm long; anthers up to 1 mm
                       long_____13. *P. palustris*
         7. All nerves of the lemma pubescent.
            12. First glume 2.5–3.5 mm long; lemma 3.5–4.5 mm
                long; blades 1–2 mm broad_____14. *P. wolfii*

12. First glume 1.5–2.5 (–2.7) mm long; lemma 1.5–3.5 mm long; blades (1–) 2–5 mm broad.

    13. Tufted perennial; inflorescence reflexed or spreading, 10–20 cm long; lemmas with 5 distinct nerves_____15. *P. sylvestris*

    13. Tufted annual or rhizomatous perennial; inflorescence ascending; lemmas with 3 distinct nerves and 2 obscure nerves.

        14. Tufted annual to 30 cm tall; culms terete; anthers 0.1–0.2 mm long_____ _____2. *P. chapmaniana*

        14. Rhizomatous perennial to 70 cm tall; culms compressed; anthers about 1 mm long___ _____7. *P. compressa*

SECTION **Annuae**

1. **Poa annua** L. Sp. Pl. 68. 1753. *Fig. 155.*

Tufted annual, sometimes rooting at the nodes, with terete culms to about 30 cm tall; sheaths loose, glabrous; blades soft, 1–3 mm broad; inflorescence 2–10 cm long, ascending; spikelets 3–6 mm long, 3- to 6-flowered; glumes narrowly ovate, obscurely nerved, with a scarious margin, the first 1.5–2.5 mm long, the second 2–3 mm long; lemmas elliptic to ovate, obtuse, thin, 5-nerved, more or less pubescent throughout on the nerves, 2.5–3.5 mm long, without a web at the base; anthers 0.8–1.0 mm long; 2n = 28 (Avdulov, 1928).

COMMON NAME: Annual Bluegrass.

HABITAT: Moist situations, particularly in waste ground.

RANGE: Native of Europe and Asia; introduced throughout North America.

ILLINOIS DISTRIBUTION: Rather common; probably in every county.

Because of the lack of a web at the base of the lemma, this species superficially resembles *Eragrostis*. In *Eragrostis*, however, the lemmas are only 3-nerved.

The Annual Bluegrass flowers in Illinois from late April until the end of the growing season. Great variation exists in the size of *P. annua*. Under favorable conditions, this species may exceed the measurements given in the description above.

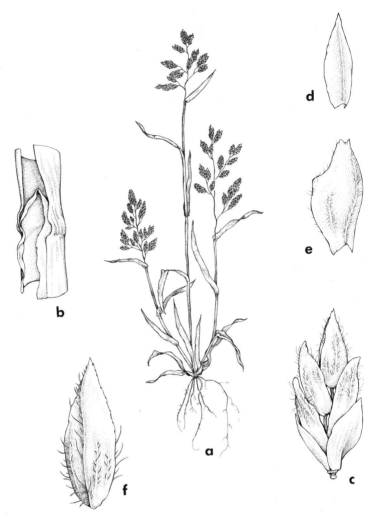

155. *Poa annua* (Annual Bluegrass). *a.* Habit, X½. *b.* Sheath, with ligule, X5. *c.* Spikelet, X7. *d.* First glume, X9. *e.* Second glume, X9. *f.* Lemma, X9.

2. Poa chapmaniana Scribn. Bull. Torrey Club 21:38. 1894.
Fig. 156.

Tufted annual with terete culms to 30 cm tall; sheaths closed; blades soft, (1–) 2–3 mm broad; inflorescence 2–8 cm long, ascending; spikelets 2.5–4.5 mm long, 3- to 6-flowered; glumes

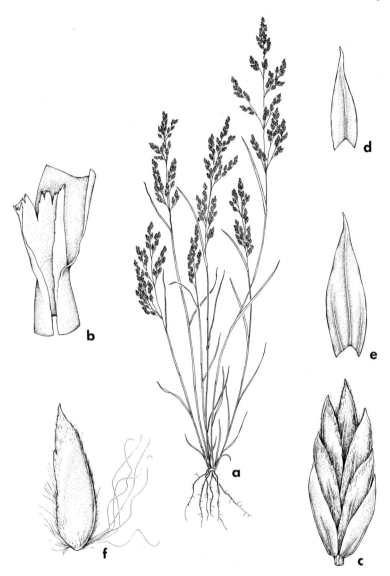

156. *Poa chapmaniana* (Annual Bluegrass). *a.* Habit, X½. *b.* Sheath, split open, to reveal ligule, X5. *c.* Spikelet, X15. *d.* First glume, X17½. *e.* Second glume, X17½. *f.* Lemma, X17½.

broadly lanceolate, obscurely nerved, with a scarious margin, the first 1.5–2.0 mm long, the second 1.5–2.5 mm long; lemmas elliptic to ovate, obtuse, thin, 1.5–2.5 mm long, distinctly 3-nerved, with 2 obscure intermediate nerves, the nerves pubescent, with a web at the base; anthers 0.1–0.2 mm long.

COMMON NAME: Annual Bluegrass.

HABITAT: Fields and waste ground.

RANGE: Delaware and Kansas, south to Texas and Florida.

ILLINOIS DISTRIBUTION: Occasional; in every county in the southern half of the state, becoming less plentiful in the northern counties.

This species flowers from mid-April until mid-September. The presence of the web on the lemma and the closed sheaths distinguish this species from *P. annua*.

SECTION **Alpinae**

3. **Poa autumnalis** Muhl. ex Ell. Bot. S. C. & Ga. 1:159. 1816. *Fig. 157.*

Poa flexuosa Muhl. Descr. Gram. 148. 1817, non Smith (1800).
Tufted perennial to 75 cm tall; ligules 1–3 mm long; blades soft, 2–3 mm broad; inflorescence 8–20 cm long, open; spikelets 5–8 mm long, 3- to 6-flowered; first glume lanceolate, 2–3 mm long, the second elliptic to ovate, 2.0–3.5 mm long; lemmas oblong, obtuse, 3.0–4.5 mm long, 5-nerved, the nerves pubescent, at least below, without a web at the base.

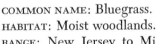

COMMON NAME: Bluegrass.

HABITAT: Moist woodlands.

RANGE: New Jersey to Michigan, south to Texas and Florida.

ILLINOIS DISTRIBUTION: Very rare; known only from Pope County (Jackson Hollow, March 28, 1963, *R. H. Mohlenbrock 11262*).

The specific epithet is misleading since this species begins to flower in late March and continues only until late June (in Illinois). This is the only perennial species of *Poa* in Illinois which lacks a cobwebby lemma.

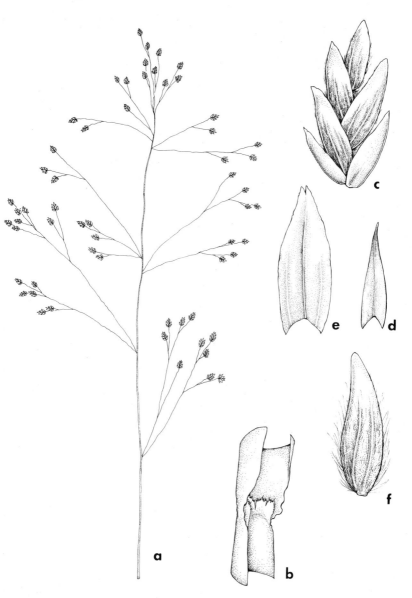

157. *Poa autumnalis* (Bluegrass).   *a.* Inflorescence, X½.   *b.* Sheath, with ligule, X5.   *c.* Spikelet, X9.   *d.* First glume, X12½.   *e.* Second glume, X12½.   *f.* Lemma, X12½.

SECTION **Pratenses**

4. **Poa arachnifera** Torr. in Marcy, Expl. Red Riv. 301. 1853.
Fig. 158.

Dioecious perennial from creeping rhizomes; culms tufted, ascending, to 75 cm tall; sheaths glabrous; blades flat, to 4 mm wide,
scabrous on the upper surface; inflorescence a dense, somewhat
contracted panicle, to 12 cm long; staminate spikelets 5- to 10-
flowered, the lemmas to 6 mm long, sparsely cobwebby at base,
otherwise glabrous; pistillate spikelets 5- to 10-flowered, the lemmas to 6 mm long, with a copious cobwebby tuft and with dense
pubescence on the strongly compressed keel and lateral nerves.

COMMON NAME: Texas Bluegrass.
HABITAT: Roadsides and pastures (in Illinois).
RANGE: Arkansas, Kansas, Oklahoma, Texas; adventive
in Illinois, North Carolina, South Carolina, Tennessee,
Georgia, Florida, Alabama, Mississippi, and Idaho.
ILLINOIS DISTRIBUTION: Known only from Winnebago
County. Fell (1955), who reported this adventive from
Winnebago County, states that it has been introduced
either for forage or for the seeding of new road shoul-
ders.

This is the only species of *Poa* in the Illinois flora which is
dioecious. The glabrous staminate spikelets and the woolly pistillate spikelets are strikingly different in appearance.

5. **Poa pratensis** L. Sp. Pl. 1753. *Fig. 159.*

Perennial with slender creeping rhizomes; culms more or less
compressed, at least at base, to nearly 1 m tall, the base 2–3 mm
thick; sheaths glabrous; ligules 1.5–2.5 mm long; blades soft, flat,
2–6 mm broad, broader than the culm; inflorescence 5–20 cm
long, spreading or ascending; spikelets 3–6 mm long, 3- to 5-flow-
ered; glumes broadly lanceolate to ovate, straight, the first 2–3
mm long, the second 2.2–3.0 mm long; lemmas elliptic to ovate,
subacute, 2.5–3.5 mm long, 5-nerved, the marginal nerves and the
keel pubescent, the intermediate nerves glabrous, with a web at
the base; anthers 1.0–1.5 mm long; 2n = 28, 56, 70 (Avdulov,
1931).

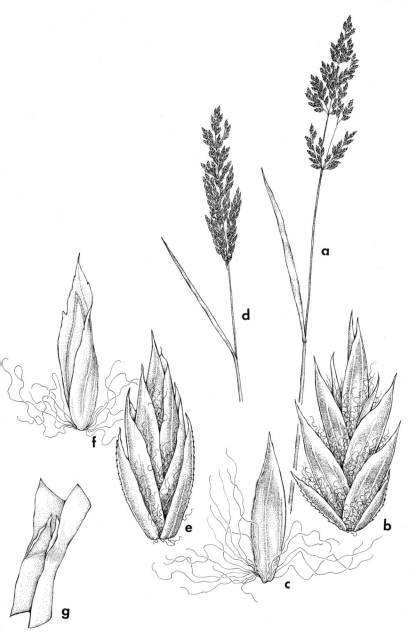

158. *Poa arachnifera* (Texas Bluegrass). *a.* Pistillate inflorescence, X7½. *b.* Pistillate spikelet, X7½. *c.* Pistillate lemma, X7½. *d.* Staminate inflorescence, X7½. *e.* Staminate spikelet, X7½. *f.* Staminate lemma, X7½. *g.* Ligule, X2½.

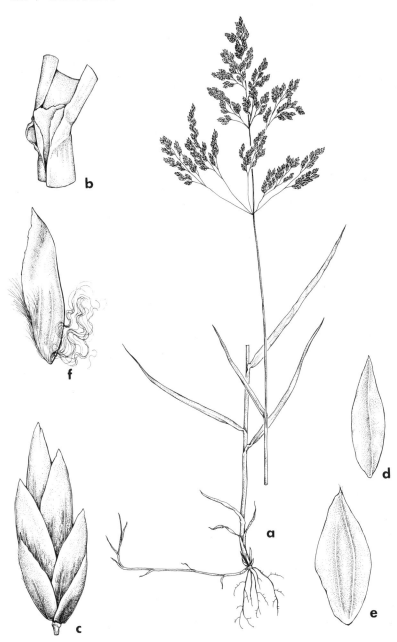

159. *Poa pratensis* (Kentucky Bluegrass). *a.* Habit, X½. *b.* Sheath, with ligule, X5. *c.* Spikelet, X10. *d.* First glume, X15. *e.* Second glume, X15. *f.* Lemma, X15.

COMMON NAME: Kentucky Bluegrass.

HABITAT: Woods, fields, and waste ground.

RANGE: Native of Europe and Asia; introduced through-out North America.

ILLINOIS DISTRIBUTION: Common; in every county.

This species flowers from mid-April to early July. It occurs in nearly all habitats and exhibits a wide range of variability in almost all characters.

Poa pratensis, along with *P. angustifolia*, is the only rhizomatous species of bluegrass in Illinois in which the intermediate nerves of the lemma are glabrous.

6. Poa angustifolia L. Sp. Pl. 67. 1753. *Fig. 160.*

Perennial with slender creeping rhizomes; culms firm, terete, less than 1 m tall, the base 1–2 mm thick; sheath glabrous; blades rather firm, the basal involute or filiform, narrower than the culm, the cauline leaves flat, 1–2 mm broad; inflorescence up to 20 cm long, spreading to ascending; spikelets 3–6 mm long, 3- to 5-flow-ered; glumes narrowly lanceolate, at least the second one arching; lemmas narrowly elliptic, subacute, 2.5–3.5 mm long, 5-nerved, the marginal nerves and the keel pubescent, the intermediate nerves glabrous, with a web at the base.

COMMON NAME: Bluegrass.

HABITAT: Woods and clearings.

RANGE: Throughout the northern half of North America, south to North Carolina, southern Illinois, Nebraska, and California.

ILLINOIS DISTRIBUTION: Known from a single collection from Union County (Southern Illinois University Pine Hills Field Station, June 20, 1968, *S. Poellot 3445*).

There is some question that *P. angustifolia* is specifically distinct from *P. pratensis*. The differences between the two, as indicated in the key, seem distinct enough. Experimental evidence is needed.

7. Poa compressa L. Sp. Pl. 69. 1753. *Fig. 161.*

Perennial with slender creeping rhizomes; culms compressed, to 70 cm tall; sheaths glabrous; blades soft, blue-green, 2–4 mm broad; inflorescence 3–15 cm long, ascending; spikelets 3–8 mm long, 3- to 8-flowered; glumes more or less lanceolate, the first 1.5–2.5 mm long, the second 1.8–2.5 mm long; lemmas elliptic to

160. *Poa angustifolia* (Bluegrass). *a.* Inflorescences, X½. *b.* Sheath, with ligule, X5. *c.* Spikelet, X5. *d.* First glume, X15. *e.* Second glume, X15. *f.* Lemma, X15.

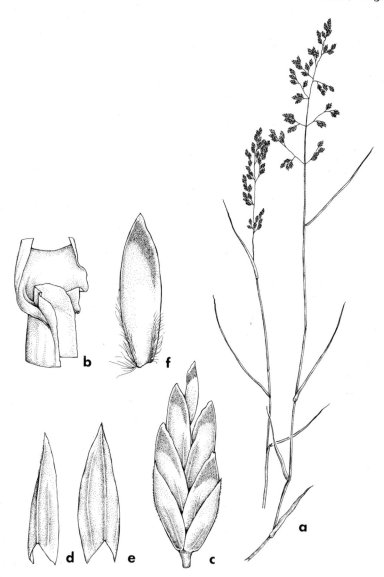

161. Poa compressa (Canadian Bluegrass). *a.* Inflorescences, X½. *b.* Sheath, with ligule, X5. *c.* Spikelet, X12½. *d.* First glume, X20. *e.* Second glume, X20. *f.* Lemma, X20.

ovate, obtuse, 2–3 mm long, obscurely nerved, pubescent on the nerves, at least below, with a sparse web at the base; anthers about 1 mm long.

COMMON NAME: Canadian Bluegrass.

HABITAT: Dry soil.

RANGE: Native of Europe and Asia; introduced throughout most of North America.

ILLINOIS DISTRIBUTION: Common; in every county.

This species is distinguished readily by the strongly compressed culms and the blue-green appearance of the leaves. It flowers from mid-May to mid-August.

SECTION **Palustres**

8. **Poa languida** Hitchc. Proc. Biol. Soc. Wash. 41:158. 1928. *Fig. 162.*

Poa debilis Torr. Fl. N. Y. 2:459. 1843, non Thuill. (1799).

Tufted perennial with weak culms to nearly 1 m tall; sheaths compressed, glabrous; ligules 1–3 mm long; blades soft, 2–5 mm broad; inflorescence 5–15 cm long, ascending to more or less nodding; spikelets 3–4 mm long, 2- to 4-flowered; glumes acute, the first 1.5–2.5 mm long, lanceolate to ovate, the second 2–3 mm long, lanceolate to elliptic; lemmas oblong, obtuse, 2.5–3.2 mm long, 5-nerved, glabrous throughout except for the web at the base; anthers 0.6–0.8 mm long.

COMMON NAME: Woodland Bluegrass.

HABITAT: Moist woodlands.

RANGE: Vermont to Minnesota, south to Iowa and Pennsylvania.

ILLINOIS DISTRIBUTION: Rare; known only from two extreme northern counties; not collected in Illinois for over 40 years. The collection by E. J. Hill from Glencoe, Cook County, near the turn of the century is the first from Illinois.

The flowering period is from June to August.

This is the only species of *Poa* in Illinois with the keel and nerves of the lemmas completely glabrous, except for the cobwebby hairs.

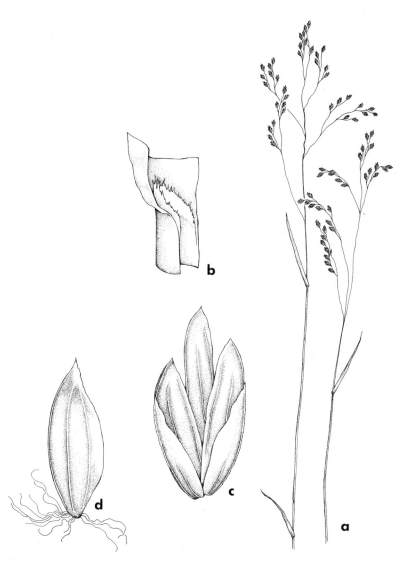

162. *Poa languida* (Woodland Bluegrass). *a.* Inflorescences, X½. *b.* Sheath, with ligule, X5. *c.* Spikelet, X12½. *d.* Lemma, X12½.

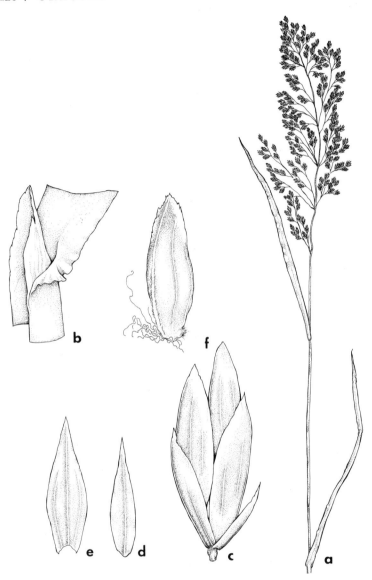

163. Poa trivialis (Meadow Grass). *a*. Inflorescence, X½. *b*. Sheath, with ligule, X5. *c*. Spikelet, X12½. *d*. First glume, X17½. *e*. Second glume, X17½. *f*. Lemma, X17½.

9. Poa trivialis L. Sp. Pl. 67. 1753. *Fig. 163.*

Poa stolonifera Hall ex Muhl. Descr. Gram. 139. 1817.

Tufted perennial, sometimes decumbent at the base; culms weak, to nearly 1 m tall, scabrous below the inflorescence; sheaths scabrous; ligules (of upper leaves) 4–8 mm long; blades soft, 2–6 mm broad; inflorescence 5–20 cm long, ascending, the pedicels scabrous; spikelets 2.5–4.0 mm long, 2- to 3-flowered; glumes lanceolate, the first 1.7–3.0 mm long, the second 2.0–3.5 mm long; lemmas narrowly ovate, acute to acuminate, 2.5–3.2 mm long, 5-nerved, pubescent on the keel only, with a web at the base.

COMMON NAME: Meadow Grass.

HABITAT: Waste ground (in Illinois).

RANGE: Native of Europe; introduced throughout most of North America.

ILLINOIS DISTRIBUTION: Rare; known only from a collection from Cook County (Chicago, May 25, 1948, *E. K. Chord s.n.*) and one from Stark County (*V. H. Chase* in 1907).

This species has the longest ligules of any *Poa* in Illinois. This character particularly distinguishes this species from *P. alsodes*, which it most nearly resembles.

10. Poa alsodes Gray, Man., ed. 2. 562. 1856. *Fig. 164.*

Tufted perennial with culms to about 80 cm tall; sheaths glabrous; ligule 1 mm long; leaves soft, thin, 2–5 mm broad; inflorescence 10–25 cm long, ascending; spikelets 3–6 mm long, 2- to 3-flowered; glumes acute, the first 2–3 mm long, lanceolate, the second 2.5–3.5 mm long, narrowly ovate; lemmas lance-ovate, acute, 2.5–4.0 mm long, obscurely 5-nerved, pubescent only on the keel, webbed at the base; anthers 0.4–0.7 mm long.

COMMON NAME: Woodland Bluegrass.

HABITAT: Moist woodlands.

RANGE: Ontario to Minnesota, south to Illinois and Virginia.

ILLINOIS DISTRIBUTION: Rare; known from two southern counties.

All other specimens from Illinois called this are *P. sylvestris*.

This species flowers from May to late June. This and *P. trivialis* are the only species in Illinois with a pubes-

164. Poa alsodes (Woodland Bluegrass). *a.* Inflorescence, X½. *b.* Sheath, with ligule, X16. *c.* Spikelet, X17½. *d.* Lemma and palea, X17½.

cent keel and glabrous nerves on the lemmas. From *P. trivialis* it is distinguished by its very short ligules and its obscurely nerved lemmas.

11. Poa paludigena Fern. & Wieg. Rhodora 20:126. 1918. *Fig. 165.*

Perennial with culms solitary or in small tufts, weak, compressed, to 60 cm tall; sheaths scabrous; ligules 0.5–1.5 mm long; blades soft, thin, 1–2 mm broad; inflorescence 3–15 cm long, widely spreading; spikelets 3–6 mm long, 2- to 5-flowered; glumes lanceolate, with a scarious margin, the first 1.7–2.2 mm long, the second 2–3 mm long; lemmas lanceolate to narrowly ovate, acute, 2.5–3.5 mm long, with 3 distinct, pubescent nerves and 2 obscure, glabrous nerves, webbed at the base; anthers 0.5–1.0 mm long.

COMMON NAME: Marsh Bluegrass.
HABITAT: Bogs.
RANGE: New York to Wisconsin, south to Illinois and Pennsylvania.
ILLINOIS DISTRIBUTION: Rare; known only from Kane County.
In Illinois, this species flowers in June and early July. Throughout its entire range, this species is very rare and, as a result, imperfectly known.
Poa paludigena resembles *P. nemoralis* and *P. palustris*, but has very lax, usually solitary culms, scabrous sheaths, and paired panicle branches near the base of the inflorescence.

12. Poa nemoralis L. Sp. Pl. 69. 1753. *Fig. 166.*

Tufted perennial with slender, terete culms to about 80 cm tall; ligules 0.5–1.0 mm long; blades 1–3 mm broad; inflorescence 5–20 cm long, open and lax; spikelets 3–6 mm long, 2- to 4-flowered; glumes narrowly lanceolate, long-acuminate, the first 2–3 mm long, the second 2.5–3.5 mm long; lemmas broadly lanceolate, acute, 2–3 mm long, with 3 distinct, pubescent nerves and 2 obscure, glabrous nerves, sparsely webbed at the base; anthers 1.2–1.6 mm long; 2n = 28 (Avdulov, 1931) 42 (Armstrong, 1937).

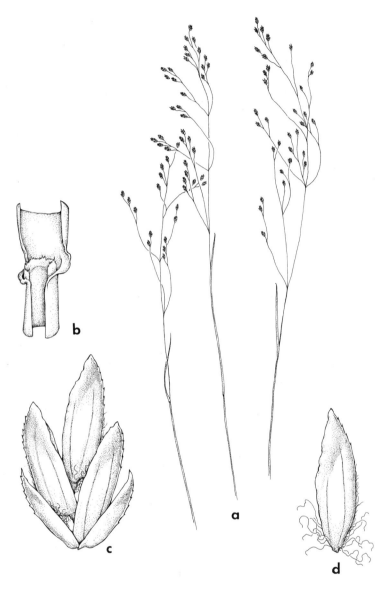

165. *Poa paludigena* (Marsh Bluegrass). *a.* Upper part of plants, X½. *b.* Sheath, with ligule, X5. *c.* Spikelet, X15. *d.* Lemma, X15.

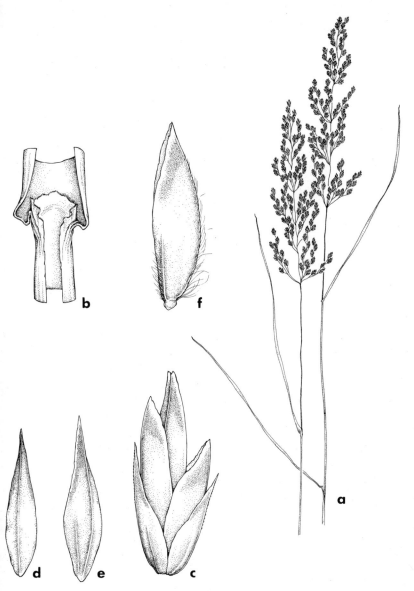

166. *Poa nemoralis* (Woodland Bluegrass). *a.* Inflorescences, X½. *b.* Sheath, with ligule, X6. *c.* Spikelet, X12½. *d.* First glume, X15. *e.* Second glume, X15. *f.* Lemma, X15.

167. *Poa palustris* (Fowl Bluegrass). *a.* Habit, X½. *b.* Sheath, with ligule, X6. *c.* Spikelet, X12½. *d.* First glume, X17½. *e.* Second glume, X17½. *f.* Lemma, X17½.

COMMON NAME: Woodland Bluegrass.

HABITAT: Open woodlands, waste ground.

RANGE: Native of Europe; introduced from Greenland to Alaska, south to Nebraska, Illinois, and Pennsylvania.

ILLINOIS DISTRIBUTION: Rare; only two collections, and neither since 1880.

Some confusion between this species and *P. palustris* is likely to be encountered, but *P. nemoralis* generally has much shorter ligules and slightly longer anthers.

13. Poa palustris L. Syst. Nat. ed. 10, 2:874. 1759. *Fig. 167.*

Poa serotina Ehrh. Beitr. Naturk. 6:83. 1791.

Poa triflora Gilib. Exerc. Phyt. 2:531. 1792.

Stout, tufted perennial, sometimes rooting at the lower nodes, to over 1 m tall; sheaths more or less loose; ligules 2–5 mm long; blades 1–3 mm broad; inflorescence 10–30 cm long, nodding; spikelets 3.0–4.5 mm long, 2- to 4-flowered; glumes ovate-lanceolate, acute, the first 2.0–2.5 mm long, the second 2–3 mm long; lemmas broadly lanceolate, acute, 2.0–3.5 mm long, with 3 distinct, pubescent nerves and 2 obscure, glabrous nerves, webbed at the base; anthers 0.8–1.0 mm long; 2n = 28 (Armstrong, 1937).

COMMON NAME: Fowl Bluegrass.

HABITAT: Wet soil.

RANGE: Newfoundland to Alaska, south to California, Illinois, and North Carolina; Europe; Asia; Africa.

ILLINOIS DISTRIBUTION: Occasional in the northern half of the state; rare in the southern half.

This species flowers from late June to early September. It is rather difficult to distinguish from *P. nemoralis*, but possesses longer ligules and somewhat smaller anthers.

Along with *P. paludigena*, it occupies more moist situations than any other species of *Poa* in Illinois.

14. Poa wolfii Scribn. Bull. Torrey Club 21:228. 1894. *Fig. 168.*

Tufted perennial with slender culms to 75 cm tall; blades soft, 1–2 mm broad; inflorescence 8–15 cm long, ascending or more or less nodding; spikelets 4–6 mm long, 2- to 4-flowered; glumes narrowly ovate, obtuse, the first 2.5–3.5 mm long, the second 3–4 mm long; lemmas narrowly ovate, acute, 3.5–4.5 mm long, 5-nerved,

villous on the nerves and keel, webbed at the base; anthers 0.8–
1.4 mm long; 2n =28 (Brown, 1939).

168. Poa wolfii (Meadow Bluegrass). *a.* Inflorescences, X½. *b.* Sheath,
with ligule, X6. *c.* Spikelet, X11. *d.* Lemma, X11.

COMMON NAME: Meadow Bluegrass.

HABITAT: Meadows and woodlands.

RANGE: Ohio to Minnesota, south to Missouri and Virginia.

ILLINOIS DISTRIBUTION: Rare; known only from three west-central counties, and not collected since 1883.

The type was collected by John Wolf from near Canton, in Fulton County. This species blooms from late April to mid-June. It has close affinities with *P. sylvestris*, but usually has larger spikelets and narrower blades, although there may be some overlapping.

15. Poa sylvestris Gray, Man. 596. 1848. *Fig. 169.*

Tufted perennial with culms more or less compressed, to about 1 m tall; ligules 1–2 mm long; blades soft, 3–5 mm broad; inflorescence 10–20 cm long, spreading or reflexed; spikelets 2.5–4.0 mm long, 2- to 5-flowered; glumes acute, with a scarious margin, the first 1.5–2.5 (–2.7) mm long, lanceolate, the second 2.0–3.5 mm long, oblong; lemmas broadly lanceolate, obtuse, 2.0–3.5 mm long, 5-nerved, villous on the nerves, webbed at the base; anthers over 1.5 mm long; 2n = 28 (Brown, 1939).

COMMON NAME: Woodland Bluegrass.

HABITAT: Moist woodlands.

RANGE: New York to Minnesota, south to Texas and Florida.

ILLINOIS DISTRIBUTION: Occasional throughout the state. This is the most common of the woodland bluegrasses in Illinois.

It flowers from early May to mid-July. It and *P. wolfii* are the only tufted perennials with all nerves of the lemma pubescent.

7. *Briza* L. – Quaking Grass

Annuals, decumbent at the base; blades flat; inflorescence a panicle; spikelets many-flowered, disarticulating above the glumes; glumes 2, nearly equal, papery, shorter than the spikelets; lemmas broad, papery, several-nerved, awnless.

The common name comes from the capillary pedicels which enable the spikelets to quake in the wind.

Only the following species has been collected in Illinois.

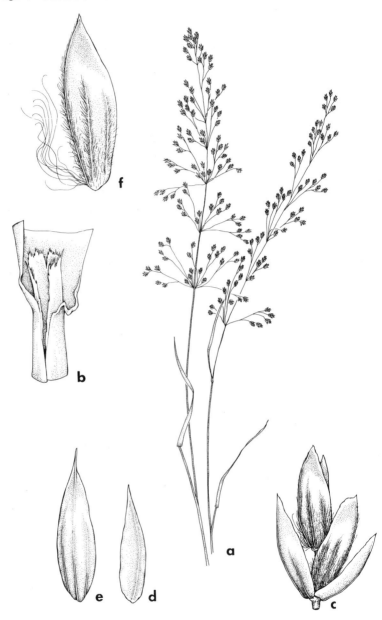

169. Poa sylvestris (Woodland Bluegrass). *a.* Inflorescences, X½. *b.* Sheath, with ligule, X6. *c.* Spikelet, X10. *d.* First glume, X12½. *e.* Second glume, X12½. *f.* Lemma, X12½.

1. Briza maxima L. Sp. Pl. 70. 1753. *Fig. 170.*

Culms to 45 cm tall, glabrous; blades 3–7 mm broad; inflorescence 5–15 cm long, with up to 8 spikelets; spikelets drooping, 12–20 mm long, nearly as broad, 10- to 20-flowered; glumes ovate to orbicular, obtuse to subacute, 6–10 mm long, puberulent, with a scarious margin; palea much shorter than the lemma; 2n = 14 (Kattermann, 1933).

COMMON NAME: Big Quaking Grass.

HABITAT: Escaped from cultivation.

RANG: Native of Europe; rarely introduced in the United States.

ILLINOIS DISTRIBUTION: Two collections, both before 1880, although the collection from Cook County is based on a cultivated specimen.

This handsome species flowers during June and July.

8. *Dactylis* L. – Orchard Grass

Cespitose perennials; sheaths compressed; blades flat; inflorescence a 1-sided panicle of fascicled spikelets; spikelets few-flowered, compressed, disarticulating above the glumes; glumes 2, unequal, keeled, shorter than the spikelets; lemmas keeled, compressed, 5-nerved, awn-tipped.

The spikelets, which are crowded to one side, distinguish this genus.

Only the following adventive species occurs in Illinois.

1. Dactylis glomerata L. Sp. Pl. 71. 1753. *Fig. 171.*

Tufted perennial with scabrous, glaucous culms to 1.2 m tall; sheaths compressed, scaberulous; ligules 5–7 mm long; blades scabrous, 2–8 mm broad; inflorescence 5–20 cm long, with a few stiff branches naked below; spikelets 3- to 6-flowered, crowded; glumes lanceolate, acuminate, ciliate along the keel, the first 4–6 mm long, the second 4.5–7.5 mm long; lemmas broadly lanceolate, ciliate along the keel, 5–8 mm long, mucronate or with an awn to 2 mm long; 2n = 42 (Hansen & Hill, 1953).

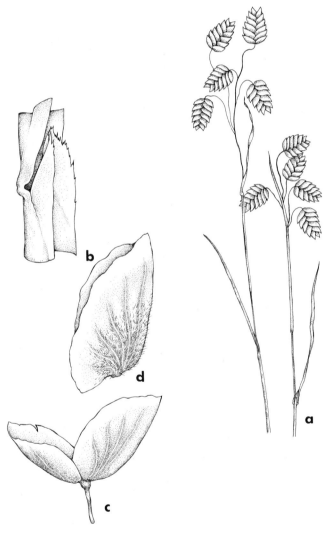

170. *Briza maxima* (Big Quaking Grass). *a.* Inflorescences, X½. *b.* Sheath, with ligule, X5. *c.* Glumes, X3. *d.* Lemma, X3.

171. *Dactylis glomerata* (Orchard Grass). *a.* Inflorescence and leaf, X½.
b. Sheath, with ligule, X2½. *c.* Spikelet, X4. *d.* Second glume, X6. *e.*
Lemma, X6.

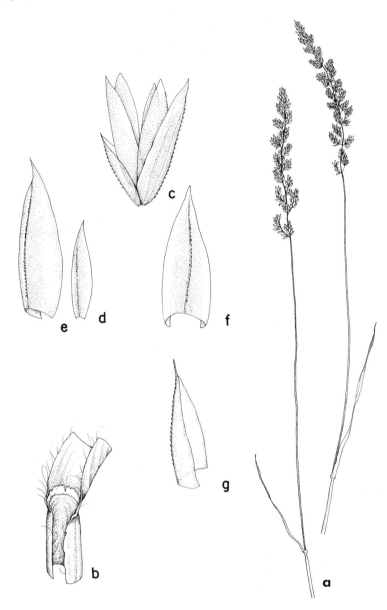

172. *Koeleria macrantha* (June Grass). *a.* Inflorescences, X½. *b.* Sheath, with ligule, X5. *c.* Spikelet, X6. *d.* First glume, X7. *e.* Second glume, X7. *f.* Lemma without awn, X7. *g.* Lemma with awn, X7.

COMMON NAME: Orchard Grass.

HABITAT: Waste ground.

RANGE: Native of Europe; introduced throughout North America.

ILLINOIS DISTRIBUTION: Common; in every county.

Some variation may be observed in the amount of pubescence on the glumes and lemmas.

This common weed flowers from early May to early July.

Tribe *Aveneae*

Annuals or perennials; inflorescence paniculate or racemose; spikelets 1- to several-flowered; glumes usually unequal, the second longer than the lowest lemma; lemmas 3- to several-nerved, awned or awnless.

The Aveneae contains more genera (18) than any other tribe of grasses in Illinois. In the newer system of classification followed here, the Aveneae includes not only the traditional genera usually assigned to it, but also much of tribe Agrostideae and all of tribe Phalarideae.

9. *Koeleria* PERS. – June Grass

Perennials (in Illinois); blades flat, narrow; inflorescence paniculate, spike-like; spikelets 2- to 4-flowered, disarticulating above the glumes; glumes somewhat unequal, slightly shorter than the spikelets; lemmas obscurely nerved, rounded on the back, awned or awnless.

This genus, like *Sphenopholis*, has glumes usually shorter than the spikelets. However, the spikelets in *Koeleria* disarticulate above the glumes, and the lemmas are sometimes short-awned. Only rarely are the lemmas awned in *Sphenopholis*.

Only the following species occurs in Illinois.

1. **Koeleria macrantha** (Ledeb.) Spreng. Mant. 2:345–46. 1924. *Fig. 172.*

Koeleria cristata Pers. Syn. Pl. 1:97. 1805, *nomen illeg.*
Koeleria gracilis Pers. Syn. Pl. 1:97. 1805, *nomen illeg.*
Aira macrantha Ledeb. Mem. Acad. St. Petersb. 5:515. 1812.
Koeleria nitida Nutt. Gen. Pl. 1:74. 1818.

Tufted perennial; culms to 60 cm tall, puberulent below the inflorescence; lower sheaths retrorsely pubescent, the upper pubescent or glabrate; blades flat to involute, glabrous or pubescent,

1–3 mm broad; inflorescence paniculate, spike-like, dense, to 12 cm long; spikelets 2- to 4-flowered, 4–6 mm long; glumes scaberulous, the first lance-oblong, 2.5–4.0 mm long, 1-nerved, the second oblong, 3–5 mm long, 3- to 5-nerved; lemmas scaberulous, obscurely 5-nerved, 3–5 mm long, awnless or with a short awn to 1 mm long; $2n = 28$ (Stebbins & Löve, 1941).

COMMON NAME: June Grass.

HABITAT: Prairies; sandy black oak woods.

RANGE: Quebec to British Columbia, south to California, Texas, Louisiana, and Delaware; Mexico.

ILLINOIS DISTRIBUTION: Occasional throughout the state. Specimens from Illinois may have either awnless or short-awned lemmas. In a few specimens, the blades may be as narrow as 1 mm, although they generally range up to 3 mm broad.

10. *Sphenopholis* SCRIBN. – Wedge Grass

Perennials; blades flat; inflorescence paniculate, narrow; spikelets 2- to 3-flowered, disarticulating below the glumes; glumes strongly unequal, rather obscurely nerved, keeled; lemmas obscurely nerved, more or less rounded on the back, usually awnless.

The glumes, which are shorter than the spikelets, give the members of this genus an appearance of some grasses in tribe Festuceae. Disarticulation of the spikelets is below the glumes, however. *Sphenopholis* differs from *Koeleria* by its glabrous axes of the inflorescence.

The three taxa of *Sphenopholis* in Illinois are all native members of the Illinois flora.

Scribner has a discussion of the genus in 1906, while Erdman (1965) has presented a comprehensive treatment of it.

KEY TO THE SPECIES OF Sphenopholis IN ILLINOIS

1. First glume subulate, less than 0.5 mm broad; lemmas smooth to scabrous; anthers less than 1 mm long_____1. S. *obtusata*
1. First glume narrowly oblong, at least 0.5 mm broad; second lemma scabrous near the apex; anthers 1.0–1.5 mm long_____2. S. *nitida*

    1. Sphenopholis obtusata (Michx.) Scribn. Rhodora 8:144. 1906.

Cespitose perennial to 1 m tall; sheaths glabrous, scabrous, or pubescent; blades 2–8 mm broad, glabrous, scabrous, or pubescent; panicle 5–25 cm long, narrow and spike-like or lobed to loose and open; spikelets 1.5–5.0 mm long; glumes glabrous, the first subulate, 1–4 mm long, 0.1–0.4 mm broad, 1-nerved, the second broadly obovate, 1.2–4.2 mm long, conspicuously or obscurely 3- to 5-nerved, firm or scarious, rounded, truncate, acute or apiculate at the apex; lemmas glabrous, the lower 1.5–4.5 mm long, smooth to scabrous.

Two generally distinctive varieties may be recognized.

1. Second glume firm, conspicuously 3- to 5-nerved, rounded to truncate at the apex; panicle dense, spike-like_____
_____1a. *S. obtusata* var. **obtusata**
1. Second glume scarious, obscurely nerved, acute or apiculate at the apex; panicle open_____1b. *S. obtusata* var. *major*

1a. Sphenopholis obtusata (Michx.) Scribn. var. **obtusata**
Fig. 173.

Aira obtusata Michx. Fl. Bor. Am. 1:62. 1803.

Aira truncata Muhl. Descr. Gram. 83. 1817.

Koeleria truncata (Muhl.) Torr. Fl. N. & Midl. U. S. 1:116. 1823.

Trisetum lobatum Trin. Mem. Acad. St. Petersb. VI. Math. Phys. Nat. 1:66. 1830.

Koeleria obtusata (Michx.) Trin. ex Steud. Nom. Bot., ed. 2, 1:849. 1840.

Koeleria lobata (Trin.) Trin. ex Steud. Nom. Bot., ed. 2, 1:849. 1840.

Eatonia obtusata (Michx.) Gray, Man. 591. 1848.

Eatonia pubescens Scribn. & Merr. Circ. U.S.D.A. Div. Agrost. 27:6. 1900.

Sphenopholis obtusata lobata (Trin.) Scribn. Rhodora 8:144. 1906.

Sphenopholis obtusata pubescens (Scribn. & Merr.) Scribn. Rhodora 8:144. 1906.

Sphenopholis pubescens (Scribn. & Merr.) Heller, Muhlenbergia 6:12. 1910.

Panicle dense and spike-like; spikelets up to 3.6 mm long; first glume 0.1–0.4 mm broad; second glume firm, scabrous, rounded or truncate at the apex; lemmas usually scabrous; 2n = 14 (Erdman, 1965).

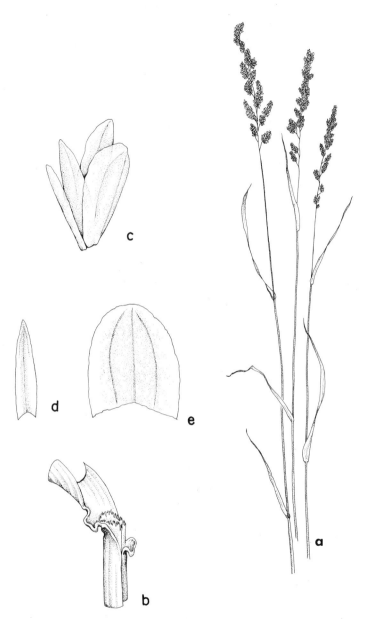

173. *Sphenopholis obtusata* var. *obtusata* (Wedge Grass). *a.* Inflorescences, X½. *b.* Sheath, with ligule, X4. *c.* Spikelet, X10. *d.* First glume, X10. *e.* Second glume, X10.

Several variations have been recorded for this taxon. Specimens exist in Illinois which have panicles more or less lobed. Other specimens show variation in the pubescence (or lack of it) of the blades and sheaths. In Illinois specimens, the demarcation of these variations is not well defined; thus, none is recognized here.

COMMON NAME: Wedge Grass.
HABITAT: Woods and prairies.
RANGE: Maine to British Columbia, south to California, Texas, and Florida; Mexico; West Indies.
ILLINOIS DISTRIBUTION: Occasional; throughout the state. This taxon flowers from May to mid-July.

1b. Sphenopholis obtusata (Michx.) Scribn. var **major** (Torr.) Erdman, Iowa State Journ. Sci. 39:310. 1965. *Fig. 174.*

Koeleria truncata var. *major* Torr. Fl. N. & Midl. U. S. 1:117. 1823.

Eatonia pennsylvanica var. *major* (Torr.) Gray, Man., ed. 2, 558. 1856.

Eatonia intermedia Rydb. Bull. Torrey Club 32:602. 1905.

Sphenopholis intermedia (Rydb.) Rydb. Bull. Torrey Club 36:533. 1909.

Panicle loose; spikelets 3–5 mm long; first glume 1–4 mm long, 0.1–0.3 mm broad; second glume scarious, acute to apiculate at the apex; lemmas rarely scabrous; 2n = 14 (Erdman, 1965).

COMMON NAME: Wedge Grass.
HABITAT: Moist woods, moist prairies.
RANGE: Newfoundland to Alaska, south to Arizona and Florida.
ILLINOIS DISTRIBUTION: Occasional; throughout the state. This taxon is nearly as abundant as *S. obtusata* var. *obtusata* and flowers during the same period. It differs in its slightly longer spikelets and its obscurely nerved, pointed second glume.

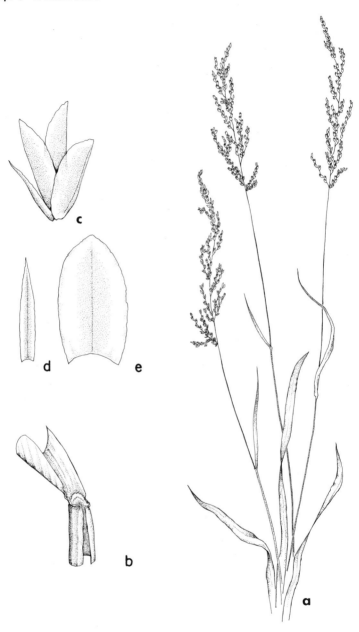

174. Sphenopholis obtusata var. *major* (Wedge Grass). *a.* Upper part of plant, X½. *b.* Sheath, with ligule, X4. *c.* Spikelet, X7½. *d.* First glume, X10. *e.* Second glume, X10.

2. Sphenopholis nitida (Bieler) Scribn. in Fern. Rhodora 47: 198. 1945. *Fig. 175.*

Aira nitida Bieler, Pl. Nov. Herb. Spreng. Cent. 8. 1807.
Aira pennsylvanica Spreng. Mem. Acad. St. Petersb. 2:299. 1807–8.

Cespitose perennial to 75 cm tall; sheaths pubescent; blades 2–5 mm broad, scabrous or pubescent; panicles 3–17 cm long, lobed; spikelets 2.8–3.2 mm long; glumes glabrous, the first narrowly oblong, 2.2–3.0 mm long, 1-nerved, the second obovate, 2.2–3.0 mm long, obscurely nerved, rounded and apiculate at the apex; lemmas 2.3–3.0 mm long, the second one scabrous near the tip; 2n = 14 (Erdman, 1965).

COMMON NAME: Shining Wedge Grass.
HABITAT: Dry woods, prairies.
RANGE: Ontario to Michigan, south to Texas and Florida.
ILLINOIS DISTRIBUTION: Not common; confined to the central and southern parts of the state; also Boone and Winnebago counties.

The broader second glume and the scabrous second lemma distinguish this species from the other taxa of *Sphenopholis* in Illinois. This species, much rarer than the other taxa, flowers from mid-May to early July.

11. *Aira* L. – Hairgrass

Annuals; blades filiform; inflorescence paniculate; spikelets 2-flowered, disarticulating above the glumes; glumes subequal, longer than the spikelets; lemmas rounded on the back, obscurely nerved, toothed and awned at the apex.

Only the following species occurs in Illinois.

1. Aira caryophyllea L. Sp. Pl. 66. 1753. *Fig. 176.*

Delicate annual with usually solitary, glabrous culms to 25 cm tall; sheaths scaberulous; blades filiform, scaberulous; panicle 3.5–7.5 cm long, more or less open; spikelets 2.2–3.0 mm long; glumes ovate, glabrous, obscurely 1- to 3-nerved, both 2.2–3.0 mm long; lemmas firm, 1.8–2.5 mm long, awned from below the middle; awn nearly straight, 2.5–3.5 mm long; 2n = 14 (Wulff, 1937).

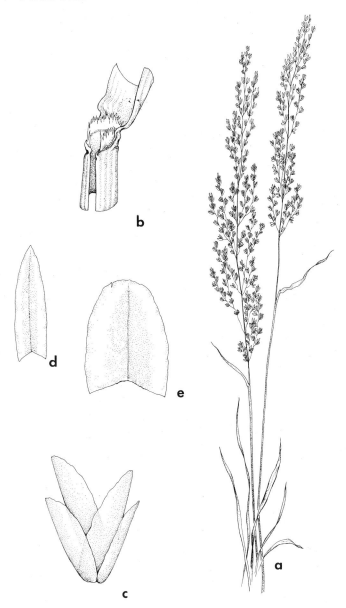

175. *Sphenopholis nitida* (Shining Wedge Grass). *a.* Upper part of plant, X½. *b.* Sheath, with ligule, X4. *c.* Spikelet, X10. *d.* First glume, X12½. *e.* Second glume, X12½.

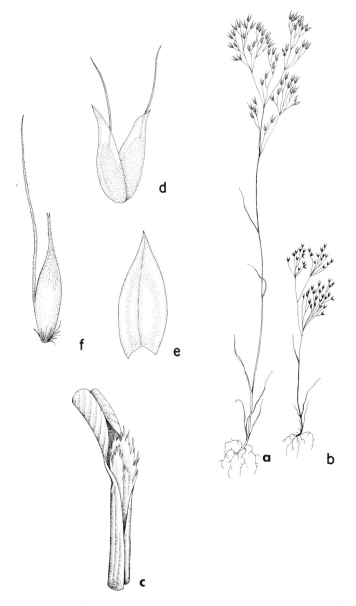

176. Aira caryophyllea (Slender Hairgrass).   *a.* Habit, X1.   *b.* Habit, X½.
c. Sheath, with ligule, X6.   *d.* Spikelet, X10.   *e.* Glume, X12½.   *f.*
Lemma, X12½.

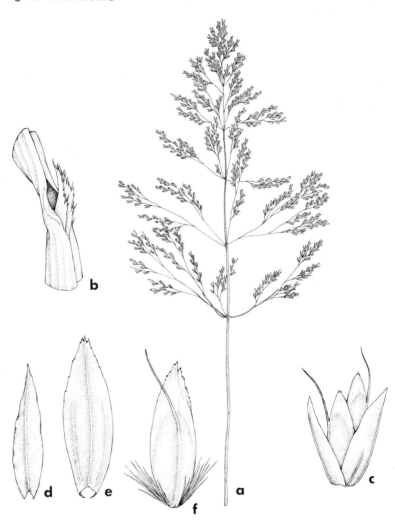

177. Deschampsia cespitosa var. *glauca* (Tufted Hairgrass). *a.* Inflorescence, X½. *b.* Sheath, with ligule, X5. *c.* Spikelet, X7½. *d.* First glume, X10. *e.* Second glume, X10. *f.* Lemma, X10.

COMMON NAME: Slender Hairgrass; Silver Hairgrass.

HABITAT: Waste ground.

RANGE: Native of Europe; adventive throughout the United States, but usually not common.

ILLINOIS DISTRIBUTION: Known only from the edge of a meadow in Piatt County (Allerton Park, May 30, 1950, *H. E. Ahles 2429*).

This species flowers from mid-May to mid-June.

12. Deschampsia BEAUV. – Hairgrass

Perennials; blades (in Illinois specimens) flat or plicate; inflorescence paniculate; spikelets 2-flowered, disarticulating above the glumes; glumes nearly equal, usually about as long as the spikelet; lemmas obscurely nerved, rounded on the back, awned, the callus bearded.

Only the following species occurs in Illinois.

1. **Deschampsia cespitosa** (L.) Beauv. var. **glauca** (Hartm.) Lindm. f. Svensk. Fan. 81. 1918. *Fig. 177.*

Deschampsia glauca Hartm. Handb. Skand. Fl. 448. 1820.
Cespitose perennial to about 1 m tall; sheaths glabrous; blades flat or plicate, scabrous above, more or less glabrous below, 1–5 mm broad; panicles 3–20 cm long, more or less open; spikelets 3–5 mm long; glumes glabrous, acute, the first 2.5–4.5 mm long, the second 3–5 mm long; lemmas glabrous, except for the bearded callus, obscurely 5-nerved, 3–5 mm long, awned; awn more or less straight, 3–6 mm long, arising from near the middle or base of the lemma; 2n = 26 (Lawrence, 1945).

COMMON NAME: Tufted Hairgrass.

HABITAT: Along creeks and in swamps.

RANGE: Newfoundland to Alaska, south to California, Illinois, and Virginia; Europe; Asia.

ILLINOIS DISTRIBUTION: Rare; confined to extreme northeastern Illinois.

The first Illinois collection was made by Vasey from near Elgin around 1861.

Deschampsia cespitosa var. *cespitosa,* which does not occur in Illinois, is a more robust plant with broader blades and somewhat larger spikelets.

Variety *glauca* flowers in Illinois from late June to mid-July.

13. Avena L. – Oats

Annuals (in Illinois); blades flat; inflorescence paniculate, open; spikelets large, 2- to 3-flowered, disarticulating above the glumes; glumes subequal, conspicuously nerved, as long as or longer than the lemmas; lemmas firm, bifid at the apex, rounded on the back, obscurely nerved, awned or awnless.

KEY TO THE SPECIES OF Avena IN ILLINOIS

1. Spikelet 3-flowered; lemmas pubescent, with a bent awn_____
   _____1. *A. fatua*
1. Spikelet 2-flowered; lemmas glabrous, awnless or with a straight
   awn_____2. *A. sativa*

1. Avena fatua L. Sp. Pl. 80. 1753. *Fig. 178.*

Tufted annual with glabrous culms to nearly 1 m tall; sheaths glabrous; blades scaberulous, 4–8 mm broad; panicle very lax; spikelets 2.0–2.5 cm long, drooping, 3-flowered; glumes glabrous, several-nerved, subequal, 1.7–2.0 cm long; lemmas pubescent, obscurely nerved, 1.0–1.6 cm long, awned from about the middle; awn 3–4 cm long, bent near the middle; 2n = 42 (Philp, 1933).

COMMON NAME: Wild Oats.
HABITAT: Waste ground.
RANGE: Native of Europe; occasionally introduced in the United States.
ILLINOIS DISTRIBUTION: Not common.
The flowers of this species are produced from late May to late September.
The first Illinois collection was made in 1894 by W. S. Moffatt from Naperville, DuPage County.

2. Avena sativa L. Sp. Pl. 79. 1753. *Fig. 179.*

Tufted annual usually branched from the base with glabrous culms to nearly 1 m tall; sheaths glabrous; blades scaberulous, 5–15 mm broad; panicle very lax; spikelets 2.0–2.5 cm long, drooping, 2-flowered; glumes glabrous, several-nerved, subequal, 1.5–2.5 cm long; lemmas glabrous, obscurely nerved, 1.5–2.0 cm long, awnless or with a straight awn less than 3 cm long.

178. Avena fatua (Wild Oats). *a.* Inflorescence, X½. *b.* Sheath, with ligule, X2½. *c.* Spikelet, X1¼. *d.* Lemma, X2½.

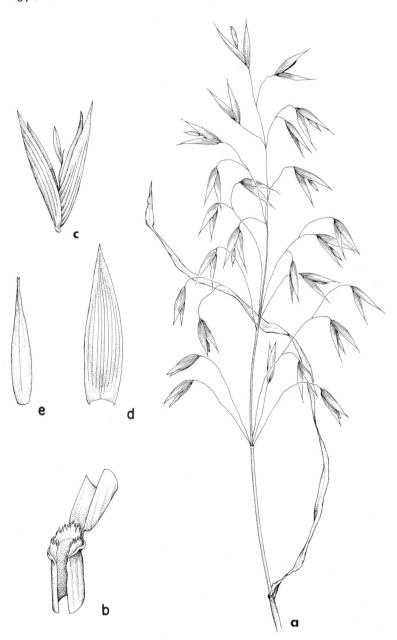

179. Avena sativa (Oats). *a.* Inflorescence, X½. *b.* Sheath, with ligule, X2½. *c.* Spikelet, X1¼. *d.* Glume, X2. *e.* Lemma, X2.

COMMON NAME: Oats.

HABITAT: Waste ground.

RANGE: Native of Eurasia; commonly cultivated in Illinois, frequently escaped, rarely if ever established.

ILLINOIS DISTRIBUTION: Throughout the state.

Some workers consider this species as a variety of *A. fatua*.

In Illinois, this species flowers from mid-May to early August.

14. Arrhenatherum BEAUV. – *Oat Grass*

Perennials; blades flat; inflorescence paniculate, narrow; spikelets 2-flowered, disarticulating above the glumes, the lower floret staminate, the upper fertile; glumes unequal, papery, nearly as long as the spikelet; lemmas rounded on the back, bearded at the base, conspicuously nerved, awnless or awned from just beneath the apex.

Only the following species occurs in Illinois.

1. Arrhenatherum elatius (L.) Presl, Fl. Cech. 17. 1819. *Fig. 180.*

Avena elatior L. Sp. Pl. 79. 1753.

Cespitose perennial with glabrous or puberulent culms to nearly 2 m tall; sheaths glabrous; blades 4–10 mm broad, scabrous; panicle purplish (in Illinois), to 30 cm long; spikelets 7–10 mm long; glumes scaberulous, ovate-lanceolate, acute to acuminate, the first 1-nerved, 4–8 mm long, the second 3-nerved, 6–10 mm long; lemmas glabrous or puberulent, 5–10 mm long, 5- to 7-nerved; awn of upper lemma none, of lower lemma to 2 cm long; $2n = 28$ (Avdulov, 1931).

COMMON NAME: Tall Oat Grass.

HABITAT: Waste ground.

RANGE: Native of Europe; occasionally introduced in the United States.

ILLINOIS DISTRIBUTION: Not common; widely scattered throughout the state.

This species has been grown often as a pasture grass in several areas of the United States. It has been known in Illinois since 1878 when Burrill collected it from Urbana, Champaign County.

The most distinguishing characteristic of this species is the

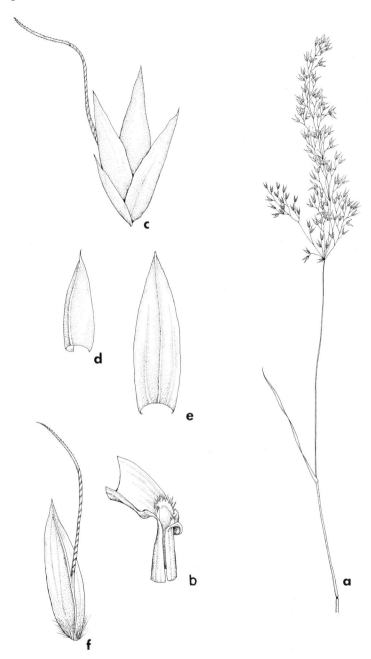

180. *Arrhenatherum elatius* (Tall Oat Grass). *a.* Inflorescence, X⅗. *b.* Sheath, with ligule, X5. *c.* Spikelet, X6. *d.* First glume, X7½. *e.* Second glume, X7½. *f.* Lemma, X7½.

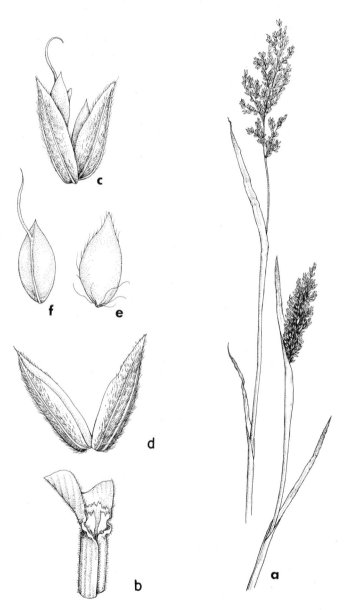

181. Holcus lanatus (Velvet Grass). *a.* Inflorescences, X½. *b.* Sheath, with ligule, X2½. *c.* Spikelet, X6. *d.* Glumes, X7. *e.* First lemma, X7½. *f.* Second lemma, X7½.

lowest floret of the 2-flowered spikelet which is only staminate.

15. *Holcus* L. – Velvet Grass

Perennials; blades flat; inflorescence paniculate, dense, contracted; spikelets 2-flowered, disarticulating below the glumes, the lowest flower perfect, the upper staminate; glumes subequal, longer than the lemmas, keeled; lemmas rounded on the back, awnless or with an awn arising beneath the apex.

Only the following species occurs in Illinois.

1. Holcus lanatus L. Sp. Pl. 1048. 1753. *Fig. 181.*

Notoholcus lanatus (L.) Nash in Britt. & Brown, Ill. Fl. 1:214. 1913.

Tufted grayish perennial with velvety culms to nearly 1 m tall; sheaths villous; blades 4–10 mm broad, villous; panicle narrow, contracted, 3–15 cm long, purplish; spikelets 3.5–5.0 mm long; glumes pubescent, acute, subequal, 4–5 mm long; lemmas glabrous, ciliate at the apex, 2.0–2.5 mm long, the lower awnless, the upper with a bent awn 1–2 mm long; 2n = 14 (Avdulov, 1928).

COMMON NAME: Velvet Grass.

HABITAT: Waste ground.

RANGE: Native of Europe; introduced throughout the United States.

ILLINOIS DISTRIBUTION: Not common. Apparently the first Illinois collection was made in 1891 by Burrill from Urbana, Champaign County.

This species flowers during June and July.

The 2-flowered spikelets bear an upper staminate floret and a lower perfect floret.

16. *Calamagrostis* ADANS. – Reed Grass

Perennials from creeping rhizomes; blades flat or involute; inflorescence paniculate, sometimes spike-like; spikelets 1-flowered, disarticulating above the glumes; glumes subequal; lemmas shorter than the glumes, 3- or 5-nerved, awned from the back, with a bearded callus.

Studies on this genus have been made by Stebbins (1930) and Nygren (1954).

KEY TO THE SPECIES OF Calamagrostis IN ILLINOIS

1. Callus of lemma shorter than or equalling the lemma; spikelets up to 4.5 mm long; awn of lemma straight.

2. Blades flat (at least when fresh), 4–8 mm broad; panicle open, more or less nodding; glumes spreading in fruit; lemmas translucent at tip_____1. *C. canadensis*
2. Blades involute, 2–4 mm broad when unrolled; panicle contracted, spike-like, erect; glumes connivent at tip in fruit; lemmas firm throughout_____2. *C. inexpansa*
1. Callus of lemma exceeding the lemma; spikelets 5–6 mm long; awn of lemma slightly bent_____3. *C. epigeios*

1. **Calamagrostis canadensis** (Michx.) Beauv. Ess. Agrost. 15:122. 1812.

Arundo canadensis Michx. Fl. Bor. Am. 1:73. 1803.
Perennial from creeping rhizomes; culms to 1.5 m tall, glabrous or nearly so; sheaths glabrous; blades flat, 4–8 mm broad, more or less glaucous, becoming involute on drying; panicle more or less nodding, open, 5–25 cm long, purplish or greenish; spikelets 2.2–3.8 mm long; glumes subequal, lanceolate to narrowly ovate, obtuse to acute to acuminate, rounded or weakly keeled on the back, glabrous to puberulent, 1.7–3.5 mm long, the tips spreading in fruit; lemmas translucent at the erose tip, 1.5–3.0 mm long, mostly glabrous, with a straight, included awn inserted near the middle; callus of lemma usually as long as the lemma.

Two varieties may be distinguished in Illinois with difficulty:

1. Panicle loosely flowered; spikelets 2.8–3.8 mm long; glumes distinctly exceeding the lemma, acute to acuminate_____
_____1a. *C. canadensis* var. *canadensis*
1. Panicle densely flowered; spikelets 2.2–2.8 mm long; glumes nearly or quite equaled by the lemma, obtuse to acute_____
_____1b. *C. canadensis* var. *macouniana*

1a. **Calamagrostis canadensis** (Michx.) Beauv. var. **canadensis** *Fig. 182 a–f.*

Arundo agrostoides Pursh, Fl. Am. Sept. 1:86. 1814.
Calamagrostis michauxii Trin. ex Steud. Nom. Bot. 1:250. 1840.
Calamagrostis canadensis var. *typica* Stebbins, Rhodora 32:40. 1930.
Panicle loosely flowered; spikelets 2.8–3.8 mm long; glumes distinctly exceeding the lemma, acute to acuminate; 2n = 42 (Nygren, 1954).

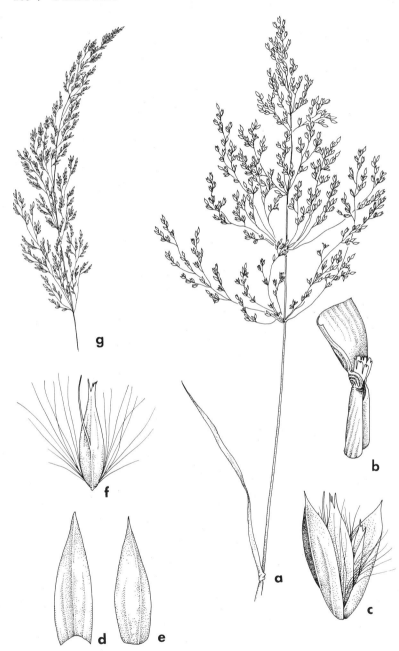

182. *Calamagrostis canadensis* (Bluejoint Grass).—var. *canadensis*. *a*. Inflorescence, X½. *b*. Sheath, with ligule, X2½. *c*. Spikelet, X12½. *d*. First glume, X12½. *e*. Second glume, X12½. *f*. Lemma, X12½.—var. *macouniana*. *g*. Inflorescence, X½.

COMMON NAME: Bluejoint Grass.

HABITAT: Moist soil.

RANGE: Greenland to Alaska, south to California, Kentucky, and North Carolina.

ILLINOIS DISTRIBUTION: Occasional in the northern half of the state, rare in the southern half, absent in the extreme south.

This tall, handsome marsh grass flowers during June and July. Although it is highly variable throughout its range, all Illinois specimens are referable to the typical variety, except for the Henry County specimen cited under the following variety.

Mead (1846) first reported this grass from Illinois as *Calamagrostis coarctata,* but this certainly is not *C. coarctata* Torr. ex Eaton.

1b. Calamagrostis canadensis (Michx.) Beauv. var. **macouniana** (Vasey) Stebbins, Rhodora 32:41. 1930. *Fig. 182g.*

Deyeuxia macouniana Vasey, Bot. Gaz. 10:297. 1885.
Calamagrostis macouniana (Vasey) Vasey, Contr. U.S. Nat. Herb. 3:81. 1892.

Panicle densely flowered; spikelets 2.2–2.8 mm long; glumes nearly or quite equaled by the lemma, obtuse to acute.

COMMON NAME: Bluejoint Grass.

HABITAT: Wet border of a railroad ditch (in Illinois).

RANGE: Prince Edward Island to New Jersey; Saskatchewan to Northwest Territory; Ohio to Minnesota, west to Washington, south to Oregon, Nebraska, Missouri, and Illinois.

ILLINOIS DISTRIBUTION: Henry County (Section 17, Geneseo Township, August 31, 1935, *R. J. Dobbs s.n.*).

In reducing this variety from species rank, Stebbins (1930) remarks that there is no sharp difference between this and typical *C. canadensis,* except that var. *macouniana* tends to have smaller spikelets and less acute glumes.

2. Calamagrostis inexpansa Gray var. **brevior** (Vasey) Stebbins, Rhodora 32:50. 1930. *Fig. 183.*

Calamagrostis stricta var. *brevior* Vasey in Rothr. in Wheeler, Rep. U. S. Survey W. 100th Merid. 6:285. 1878.

Tufted perennial from creeping rhizomes; culms to 1 m tall, scabrous beneath the panicle, otherwise glabrous; sheaths glabrous
or scabrous; blades involute, 2–4 mm broad, scabrous, glaucous;

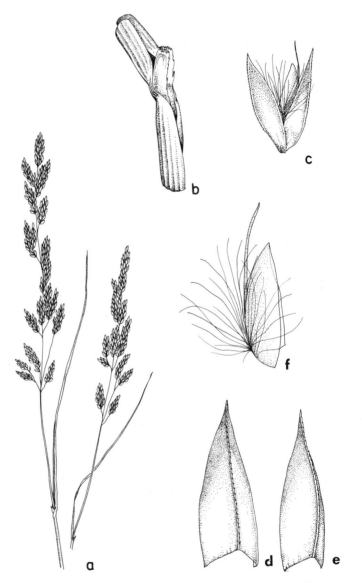

183. *Calamagrostis inexpansa* (Northern Reed Grass). *a.* Inflorescences,
X½. *b.* Sheath, with ligule, X4. *c.* Spikelet, X7½. *d.* First glume, X12½.
e. Second glume, X12½. *f.* Lemma, X12½.

panicle spike-like, erect, 5–20 cm long; spikelets 3.0–4.5 mm long; glumes subequal, lanceolate, acute to acuminate, rounded or weakly keeled on the back, scabrous, the tips connivent in fruit, the first 2.5–5.0 mm long, the second 2.5–4.8 mm long, purplish; lemmas firm, toothed at the tip, 2.5–3.5 mm long, with a straight, included awn inserted near the middle; callus of lemma shorter than the lemma; 2n = 28, 56, 84 (Nygren, 1954).

COMMON NAME: Northern Reed Grass.
HABITAT: Wet ground.
RANGE: Newfoundland to British Columbia, south to California, New Mexico, Illinois, and New York.
ILLINOIS DISTRIBUTION: Rare: known only from the extreme northern counties of Cook, Lake, JoDaviess, and Winnebago. It was collected initially in Illinois in 1873 by H. H. Babcock from Hyde Park, Chicago. Illinois specimens are compactly flowered and have somewhat smaller florets than typical var. *inexpansa*. The Illinois material may be known as var. *brevior*. It flowers during June and July.

3. **Calamagrostis epigeios** (L.) Roth, Tent. Fl. Germ. 1:34. 1788. *Fig. 184.*

Arundo epigeios L. Sp. Pl. 81. 1753.
Perennial from creeping rhizomes; culms to 1.5 m tall; blades 4–10 mm broad, flat or becoming involute on drying, scabrous; panicle erect, narrow, 25–35 cm long; spikelets 5–6 mm long; glumes subequal, linear-lanceolate, attenuate, 5–6 mm long; lemma membranous, 2–3 mm long, with a slightly bent awn about as long as the glumes and inserted near the middle of the lemma; callus of lemma much exceeding the lemma; 2n = 28, 35, 42, 56, 70 (Nygren, 1954).

COMMON NAME: Feathertop.
HABITAT: Strip-mine (in Illinois).
RANGE: Native of Europe and Asia; sparingly introduced in the United States.
ILLINOIS DISTRIBUTION: Known only from a single collection from Randolph County (strip-mine area, June 29, 1950, *A. Grandt s.n.*).

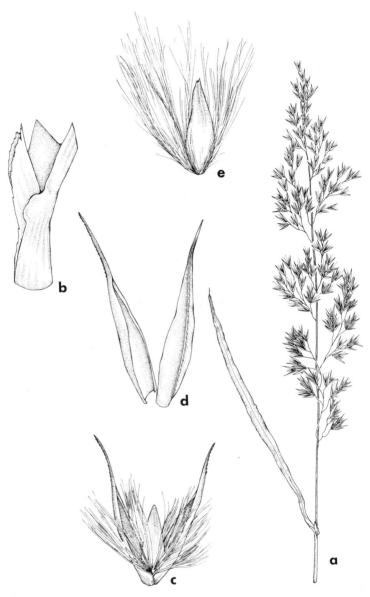

184. Calamagrostis epigeios (Feathertop). *a.* Inflorescence, X½. *b.* Sheath, with ligule, X5. *c.* Spikelet, X7½. *d.* Glumes, X10. *e.* Lemma, X10.

17. *Ammophila* HOST. – Beach Grass

Perennials from creeping rhizomes; blades involute above, flat at base; inflorescence paniculate, contracted to appear spike-like; spikelets 1-flowered, flattened, disarticulating above the glumes; glumes subequal, papery, keeled, obscurely nerved, shorter than the glumes, awnless, with a tuft of hairs on the callus.

Only the following species occurs in Illinois. It is an important sandbinding grass. It has been discussed thoroughly by Fernald (1920).

1. **Ammophila breviligulata Fern.** Rhodora 22:71. 1920. *Fig. 185.*

Coarse perennial with stiff, glabrous culms to nearly 1 m tall; sheaths glabrous; blades flat below, involute near the tip, 4–8 mm broad when unrolled, scabrous above, glabrous beneath; panicle spike-like, densely flowered, to 35 cm long, the base usually enclosed in the sheath; spikelets 8–15 mm long; glumes 9–15 mm long, linear-lanceolate, obtuse to acute, scabrous on the keel and sometimes along the sides, the first 1-nerved, the second 3-nerved; lemmas 7–14 mm long, obtuse, scabrous, obscurely 3- to 5-nerved, with the callus beard to 3 mm long; 2n = 28 (Church, 1929).

COMMON NAME: Beach Grass.
HABITAT: Sand dunes.
RANGE: Newfoundland to North Carolina; Great Lakes region.
ILLINOIS DISTRIBUTION: Confined to the sand dunes along Lake Michigan in Cook and Lake counties. The first Illinois collection was made along the lake shore in Chicago in 1860 by F. Scammon. Vasey reported it in 1861 as *Calamagrostis arenaria*.

This species flowers during July, August, and September. It is similar to species of *Calamagrostis*, from which it is distinguished by being awnless.

18. *Agrostis* L. – Bent Grass

Tufted annuals or cespitose or rhizomatous perennials; blades flat or involute; inflorescence paniculate, spreading or contracted; spikelets 1-flowered, disarticulating above the glumes; glumes subequal, more or less keeled; lemmas smaller than the glumes,

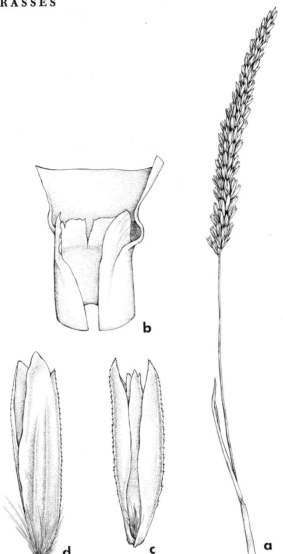

185. *Ammophila breviligulata* (Beach Grass). *a.* Inflorescence, X½. *b.* Sheath, with ligule, X7½. *c.* Spikelet, X5. *d.* Lemmas, X6.

rounded on the back, obscurely nerved (except in A. *elliottiana*), awnless (awned in A. *elliottiana*); palea sometimes absent.

KEY TO THE SPECIES OF Agrostis IN ILLINOIS

1. Annuals; lemma sharply nerved, with a flexuous awn to 10 mm long
   _____1. A. *elliottiana*

1. Perennials; lemma obscurely nerved, awnless (rarely a very short awn present in a variety of *A. scabra*).

  2. Tufted perennials without rhizomes; palea absent, or minute and nerveless.

    3. Flat blades 1–2 mm broad; spikelets 1.2–2.0 mm long; lemma 0.5–1.0 mm long_____2. *A. hyemalis*

    3. Flat blades 2–6 mm broad; spikelets 2–3 mm long; lemma 1.3–2.0 mm long.

      4. Panicle branches harshly scabrous, bearing florets only near the tip_____3. *A. scabra*

      4. Panicle branches glabrous or nearly so, bearing florets from near the middle to the tip_____4. *A. perennans*

  2. Rhizomatous or stoloniferous perennials; palea at least one-half as long as the lemma, 2-nerved.

    5. Some of the panicle branches bearing florets to base; ligule 2–6 mm long_____5. *A. alba*

    5. None of the panicle branches bearing florets to the base; ligule less than 2 mm long_____6. *A. tenuis*

1. **Agrostis elliottiana** Schult. Mantissa 2:202. 1824. *Fig. 186.*

Agrostis arachnoides Ell. Bot. S.C. & Ga. 1:134. 1816, non Poir. (1810).

Delicate, cespitose annual with erect or decumbent culms to 50 cm tall; blades flat, 1–2 mm broad; panicle spreading, purplish or green, to 25 cm long; spikelets 1.5–2.0 mm long; glumes 1.5–2.0 mm long, lanceolate, acute, more or less scabrous on the keel; lemma 1.0–1.6 mm long, sharply 5-nerved, pilose at the base, scabrous elsewhere, with 2 teeth and an awn at the apex; awn inserted beneath the apex, flexuous, to 10 mm long; palea none or minute.

COMMON NAME: Awned Bent Grass.

HABITAT: Dry soil, particularly on bluffs.

RANGE: Maryland to Kansas, south to Texas and Florida.

ILLINOIS DISTRIBUTION: Restricted to the southern one-half of Illinois; apparently most abundant along the Shawneetown Ridge.

Except for *A. scabra* var. *tuckermanii*, which has an awn about 1 mm long, this is the only species of *Agrostis* in Illinois which has an awned lemma. The flowering period for this grass is from early May to mid-July.

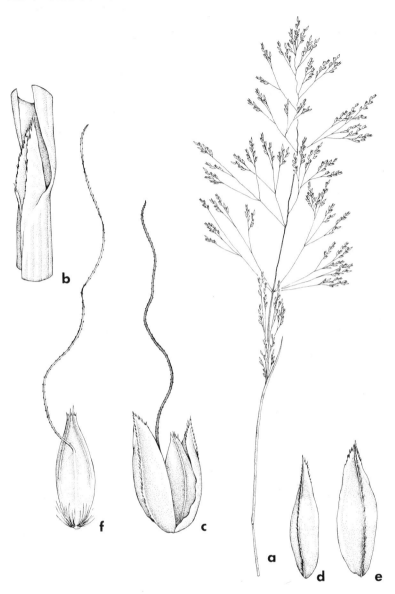

186. Agrostis elliottiana (Awned Bent Grass). *a.* Inflorescence, X½. *b.* Sheath, with ligule, X5. *c.* Spikelet, X15. *d.* First glume, X15. *e.* Second glume, X15. *f.* Lemma, X15.

2. **Agrostis hyemalis** (Walt.) BSP. Prel. Cat. N.Y. 68. 1888.
Fig. 187.

Cornucopiae hyemalis Walt. Fl. Carol. 73. 1788.
Trichodium laxiflorum Michx. Fl. Bor. Am. 1:42. 1803.
Agrostis laxiflora Poir. in Lam. Encycl. Sup. 1:255. 1810.

More or less tufted perennials without rhizomes or stolons; culms erect or decumbent, to 75 cm tall; blades flat, 1–2 mm broad, or involute; panicle purple, loose and open, to 30 cm long; spikelets 1.2–2.5 mm long, congested in terminal clusters; glumes 1.2–2.5 mm long, acute, scabrous on the keel, the tips distinct during fruiting; lemma 0.5–1.0 mm long, obtuse to subacute, awnless; palea none to minute.

COMMON NAME: Tickle Grass.
HABITAT: Woods and fields.
RANGE: Massachusetts to Minnesota and Kansas, south to Texas and Florida.
ILLINOIS DISTRIBUTION: Common throughout the state; undoubtedly in every county.
This common species has the smallest lemma of any *Agrostis* in Illinois. It flowers from March to early June. Early Illinois workers apparently did not distinguish this species from *A. scabra.*

3. **Agrostis scabra** Willd. Sp. Pl., ed. 2, 1:370. 1797.

Cespitose perennial with rhizomes or stolons; culms erect or geniculate, to 80 cm tall; blades flat, 2–5 mm broad, the basal leaves involute; panicle purple or green, loose and open, to 40 cm long; spikelets 2–3 mm long, borne only near tip of the harshly scabrous panicle branchlets; glumes lanceolate, acuminate, 2–3 mm long, scabrous on the keel, the tips connivent during fruiting; lemma 1.3–2.0 mm long, awnless or with a short awn; palea none to minute.

Two forms occur in Illinois.

1. Lemmas awnless_____3a. *A. scabra* f. *scabra*
1. Lemmas awned_____3b. *A. scabra* f. *tuckermanii*

3a. **Agrostis scabra** Willd. f. **scabra** *Fig. 188 a–f.*
Lemmas awnless.

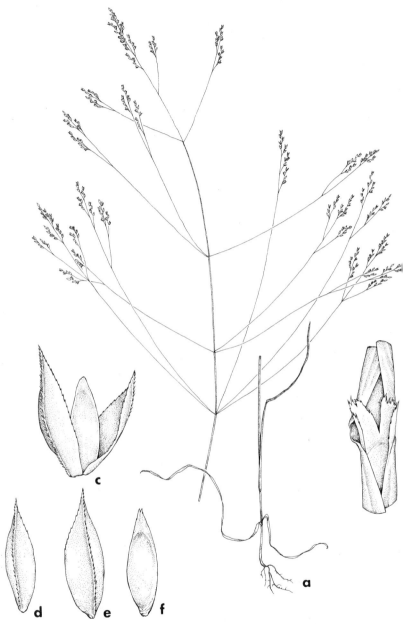

187. *Agrostis hyemalis* (Tickle Grass).   *a.* Inflorescence and base of plant, X½.   *b.* Sheath, with ligule, X5.   *c.* Spikelet, X17½.   *d.* First glume, X17½.   *e.* Second glume, X17½.   *f.* Lemma, X17½.

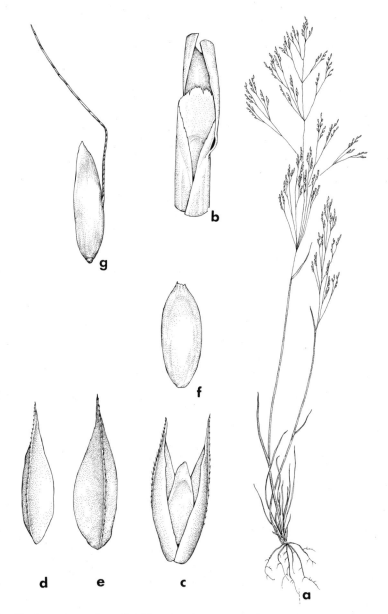

188. Agrostis scabra (Tickle Grass).—var. *scabra. a.* Habit, X½. *b.* Sheath, with ligule, X5. *c.* Spikelet, X15. *d.* First glume, X15. *e.* Second glume, X15. *f.* Lemma, X15—var. *tuckermanii. g.* Lemma, X15.

COMMON NAME: Tickle Grass.

HABITAT: Moist or dry, usually open, soil.

RANGE: Labrador to Alaska, south to California, Arizona, Illinois, and South Carolina.

ILLINOIS DISTRIBUTION: Occasional in the northern counties; absent from the southern one-fourth of the state. This species is distinguished from A. *perennans* by its harshly scabrous panicle branches, and from A. *hyemalis* by its larger spikelets and connivent glume-tips during fruiting. It is sometimes confused with A. *hyemalis*.

3b. Agrostis scabra Willd. f. **tuckermanii** Fern. Rhodora 35: 207. 1933. *Fig. 188g.*

Lemmas awned.

Known only from Cook County.

4. Agrostis perennans (Walt.) Tuckerm. Am. Jour. Sci. 45:44. 1843. *Fig. 189.*

Cornucopiae perennans Walt. Fl. Carol. 74. 1788.

Agrostis elegans Salisb. Prodr. Stirp. 25. 1796.

Agrostis michauxii Trin. Gram. Unifl. 206. 1824, non Zucc. (1809).

Agrostis scabra var. *perennans* Wood, Class-book 774. 1861.

Agrostis perennans var. *aestivalis* Vasey, Contr. U. S. Nat. Herb. 3:76. 1892.

Cespitose perennial without stolons or rhizomes; culms erect or decumbent, to nearly 1 m tall; blades flat, 3–6 mm broad; panicle green, spreading, to 30 cm long; spikelets 2–3 mm long, borne from near the middle of the more or less glabrous panicle branches; glumes 2–3 mm long, lanceolate, acuminate, scabrous on the keel; lemma 1.3–2.0 mm long, awnless; palea none or minute; $2n = 14$ (Sokolovskaja, 1938).

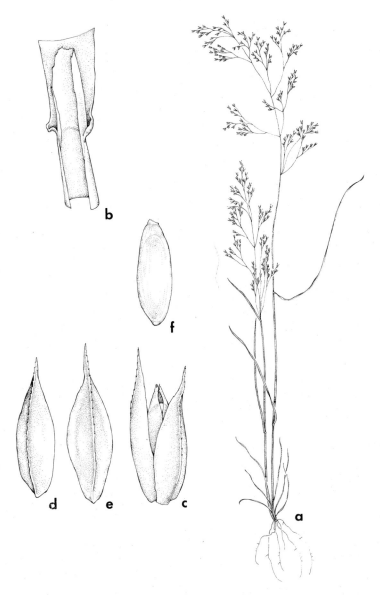

189. Agrostis perennans (Upland Bent Grass). *a.* Habit, X½. *b.* Sheath, with ligule, X5. *c.* Spikelet, X17½. *d.* First glume, X17½. *e.* Second glume, X17½. *f.* Lemma, X17½.

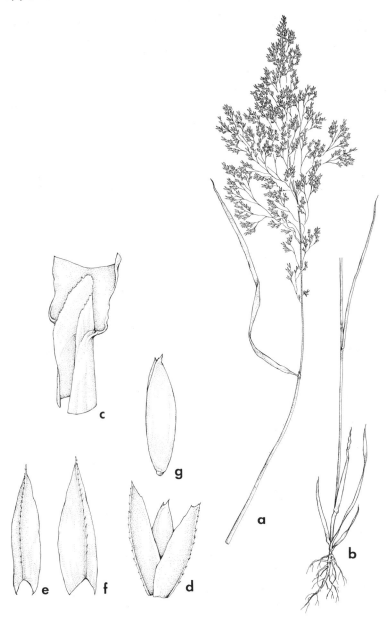

190. Agrostis alba var. *alba* (Red Top). *a.* Inflorescence, X½. *b.* Base of plant, X½. *c.* Sheath, with ligule, X5. *d.* Spikelet, X12½. *e.* First glume, X15. *f.* Second glume, X15. *g.* Lemma, X15.

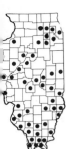

COMMON NAME: Upland Bent Grass.

HABITAT: Dry woodlands.

RANGE: Quebec to Minnesota, south to Texas and Florida.

ILLINOIS DISTRIBUTION: Common in the southern one-half of the state, becoming increasingly less common northward.

Vasey described slenderer specimens with longer panicles from Athens, Illinois, as var. *aestivalis*. The characters supposedly separating var. *aestivalis* from the typical variety intergrade freely so that it is not practical to maintain var. *aestivalis*. The Upland Bent Grass flowers from late June through September.

5. Agrostis alba L. Sp. Pl. 63. 1753.

Matted perennial with creeping rhizomes or stolons; culms erect or decumbent, to over 1 m long; blades flat or involute, to 10 mm broad; panicle purplish to straw-colored, spreading to ascending, to 30 cm long; spikelets 2.0–3.5 mm long; glumes lanceolate, acute, 2.0–3.5 mm long, scabrous on the keel; lemma acute, 1.5–3.0 mm long, awnless; palea one-half to two-thirds as long as the lemma.

Voss (1966) has given reasons why he believes that *Agrostis alba* L. should be called *A. stolonifera* L. Since there are serious biological problems as well as nomenclatural problems in this complex, I prefer to retain the commonly employed *Agrostis alba* L. until a more comprehensive study of the biological problems is made.

Two rather distinct varieties occur in Illinois. Both are treated by some authors as distinct species.

1. Blades 5–10 mm broad; rhizomes present; culms erect, rarely decumbent; panicle purple, the branches spreading_____
_____5a. A. alba var. alba
1. Blades 1–5 mm broad; stolons present; culms decumbent, rarely erect; panicle usually straw-colored, the branches ascending_____
_____5b. A. alba var. palustris

5a. Agrostis alba L. var. alba *Fig. 190.*

Agrostis dispar Michx. Fl. Bor. Am. 1:52. 1803.
Agrostis alba var. *major* Gaud. Fl. Helv. 1:189. 1828.
Agrostis alba var. *dispar* (Michx.) Wood, Class-book 774. 1861.

Agrostis stolonifera L. var. *major* (Gaud.) Farw. Rep. Mich. Acad. Sci. 21:351. 1920.

Rhizomes present; culms erect, rarely decumbent; blades 5–10 mm broad; panicle purple, the branches spreading; 2n = 28 (Sokolovskaja, 1938).

COMMON NAME: Red Top.

HABITAT: Fields.

RANGE: Newfoundland to Yukon, south to California, Texas, and Georgia.

ILLINOIS DISTRIBUTION: Common throughout the state; in every county.

This common grass flowers from June to mid-September. This variety differs from var. *palustris* by its broader blades, its erect culms with spreading panicle branches, and the presence of rhizomes.

5b. **Agrostis alba** L. var. **palustris** (Huds.) Pers. Syn. Pl. 1:76 1805. *Fig. 191.*

Agrostis palustris Huds. Fl. Angl. 27. 1762.

Agrostis polymorpha var. *palustris* Huds. Fl. Angl., ed. 2, 32. 1778.

Agrostis maritima Lam. Encycl. 1:61. 1783.

Agrostis stolonifera L. var. *compacta* Hartm. Skand. Fl. Handb., ed. 4, 24. 1843.

Agrostis stolonifera L. var. *palustris* (Huds.) Farw. Rep. Mich. Acad. Sci. 21:351. 1920.

Stolons present; culms decumbent, rarely erect; blades 1–5 mm broad; panicle usually straw-colored, the branches ascending; 2n = 28 (Church, 1936), 42 (Müntzing, 1937).

COMMON NAME: Creeping Bent Grass.

HABITAT: Wet ground.

RANGE: Greenland to British Columbia, south to California, New Mexico, Illinois, and Virginia.

ILLINOIS DISTRIBUTION: Occasional throughout the state, but apparently nowhere abundant.

This taxon sometimes is considered to be a distinct species from *A. alba*, but the existing evidence indicates its best placement to be as a variety of *A. alba*.

191. Agrostis alba var. *palustris* (Creeping Bent Grass). *a.* Inflorescence, X½. *b.* Base of plant, X½. *c.* Sheath, with ligule, X5. *d.* Spikelet, X17½.

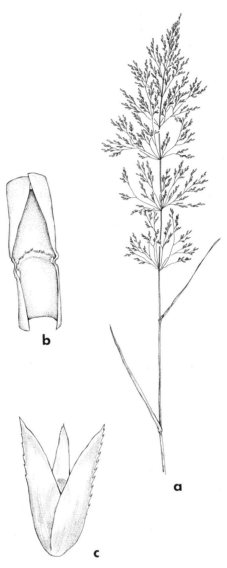

192. Agrostis tenuis (Rhode Island Bent). *a*. Inflorescence, X½. *b*. Sheath, with ligule, X5. *c*. Spikelet, X17½.

6. Agrostis tenuis Sibth. Fl. Oxon. 36. 1794. *Fig. 192.*

Agrostis capillaris Huds. Fl. Angl. 27. 1762, non L. (1753). Tufted perennial with stolons; culms to 75 cm tall; blades flat, 1–5 mm broad; panicle bronze or purple, ascending to spreading, to 20 cm long; spikelets 2–3 mm long; glumes lanceolate, acute to acuminate, 1.5–2.5 mm long, scabrous on the keel; lemma acute, 1.5–2.5 mm long, awnless; palea one-half as long as the lemma; 2n = 28 (Avdulov, 1931).

COMMON NAME: Rhode Island Bent.

HABITAT: Waste ground.

RANGE: Native of Europe; escaped from lawns in the northern United States.

ILLINOIS DISTRIBUTION: Scattered in some northern counties, but nowhere common.

The flowering period for this species is June to September. Rhode Island Bent is a popular lawn grass since it does well in full sun if properly watered, fertilized, and mowed.

19. *Cinna* L. – Wood Reed

Robust perennials; blades flat; inflorescence paniculate, large; spikelets 1-flowered, disarticulating below the glumes; glumes subequal, keeled; lemmas rounded on the back, 3-nerved, short-awned from below the apex.

KEY TO THE SPECIES OF Cinna IN ILLINOIS

1. Panicle gray-green, the branches mostly ascending; spikelets 4.0–6.5 mm long; second glume 3-nerved; awn less than 0.5 mm long__ _____1. *C. arundinacea*
1. Panicle green, the branches mostly spreading; spikelets 2.5–4.0 mm long; second glume 1-nerved; awn 0.5–1.5 mm long__2. *C. latifolia*

1. Cinna arundinacea L. Sp. Pl. 5. 1753. *Fig. 193.*

Perennial with erect, glabrous culms up to 1.5 m tall; sheaths glabrous; blades 6–15 mm broad, scabrous; panicle to 30 cm long, gray-green, the branches ascending; spikelets 4.0–6.5 mm long; glumes lanceolate, acute, scabrous on the keel, the first 3.0–5.5 mm long, the second 4.0–6.5 mm long, 3-nerved; lemma 3.5–6.0 mm long, awned from the back; awn less than 0.5 mm long.

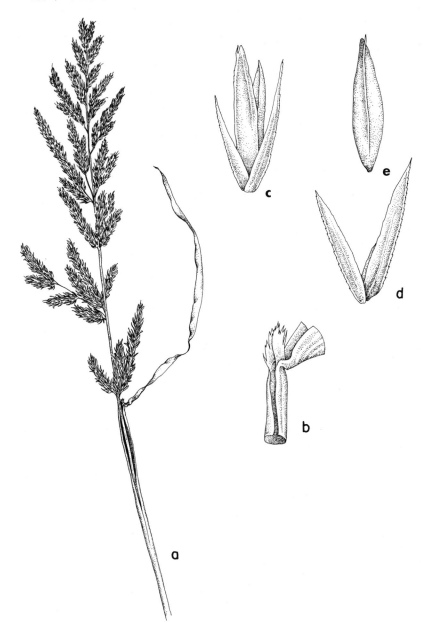

193. *Cinna arundinacea* (Stout Reed Grass). *a.* Inflorescence, X½. *b.* Sheath, with ligule, X1½. *c.* Spikelet, X7½. *d.* Glumes, X7½. *e.* Lemma, X7½.

COMMON NAME: Stout Wood Reed.

HABITAT: Moist woodlands; damp soil.

RANGE: Maine to Ontario and South Dakota, south to Texas and Georgia.

ILLINOIS DISTRIBUTION: Throughout the state, but more abundant in the southern one-half of the state.

This tall, handsome grass flowers from mid-July to late September. It differs from the following rarer species in its longer spikelets and its ascending panicle branches.

2. **Cinna latifolia** (Trev.) Griseb. in Ledeb. Fl. Ross. 4:435. 1853. *Fig. 194.*

Agrostis latifolia Trev. ex Gopp. Beschr. Bot. Gart. Breslau 82. 1830.

Cinna pendula Trin. Mem. Acad. Sci. St. Peterb. VI. Sci. Nat. 4(1):280. 1841.

Cinna arundinacea var. *pendula* (Trin.) Gray, Man., ed. 2, 545. 1856.

Perennial with erect, glabrous culms up to 1.5 m tall; sheaths glabrous; blades 7–15 mm broad, scabrous; panicle to 40 cm long, green, the branches spreading or even drooping; spikelets 2.5–4.0 mm long; glumes lanceolate, acute, scabrous on the keel, both 2.5–4.0 mm long; lemma 2.5–4.0 mm long, awned from the back; awn 0.5–1.5 mm long.

COMMON NAME: Drooping Wood Reed.

HABITAT: Moist woodlands.

RANGE: Labrador to Alaska, south to California, Illinois, and North Carolina; Europe; Asia.

ILLINOIS DISTRIBUTION; Rare; confined to the extreme northern counties of the state.

This northern grass flowers from early July to late September, usually coming into bloom a few days before *C. arundinacea.* Occasional specimens may be found in which the panicles are pendulous.

20. *Anthoxanthum* L. – Vernal Grass

Annuals or perennials; blades flat; panicles contracted, spike-like; spikelets with 1 perfect flower and two empty lemmas below the perfect one, disarticulating above the glumes; glumes unequal, 1- to 3-nerved; lemmas rounded on the back, the empty ones awned, the fertile one usually awnless.

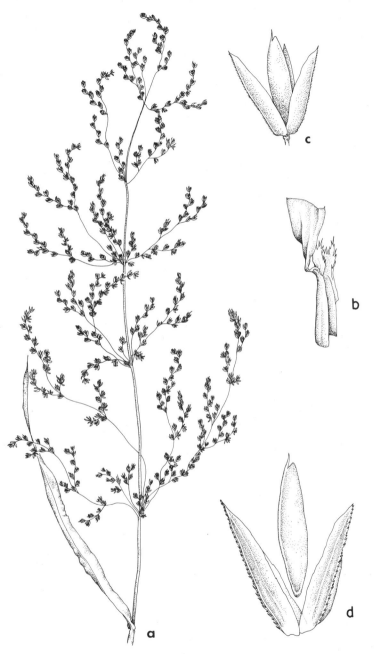

194. Cinna latifolia (Drooping Wood Reed). *a.* Inflorescence, X½. *b.* Sheath, with ligule, X2½. *c.* Spikelet, X7½. *d.* Glumes and lemma, X12½.

This genus, along with *Hierochloë* and *Phalaris,* traditionally is placed in tribe Phalarideae. Recent evidence seems to indicate that these three genera are best grouped in tribe Aveneae.

Anthoxanthum differs from *Hierochloë* in its sterile lower florets and from *Phalaris* in its awned sterile lemmas.

KEY TO THE SPECIES OF Anthoxanthum IN ILLINOIS

1. Spikelets brownish-green, 8–10 mm long; glumes pubescent; awns of empty lemmas included or barely exserted; perennials to nearly 1 m tall_____1. *A. odoratum*
1. Spikelets whitish-green, 5–7 mm long; glumes glabrous; awns of empty lemmas long-exserted; annuals to 35 cm tall__2. *A. aristatum*

1. Anthoxanthum odoratum L. Sp. Pl. 28. 1753. *Fig. 195.* Tufted perennial; culms erect, to nearly 1 m tall; blades scabrous to villous, 2–5 mm broad; panicle to 6 cm long, long-exserted; spikelets brownish-green, 8–10 mm long; glumes pubescent, the first 3.5–4.0 mm long, 1-nerved, the second 7–9 mm long, 3-nerved; fertile lemma terminal, broadly rounded, awnless, glabrous, enclosed by the sterile lemmas; sterile lemmas 3.0–3.5 mm long, golden-pubescent, the first awned below the apex, the second awned near the base, the awns included or barely exserted; 2n = 10, 20 (Oestergren, 1942).

COMMON NAME: Sweet Vernal Grass.

HABITAT: Fields.

RANGE: Native of Europe; escaped throughout the eastern United States and the Pacific Coast.

ILLINOIS DISTRIBUTION: Rare; known from four counties in extreme northeastern Illinois.

This grass has sweetly fragrant foliage similar to that of *Hierochloë odorata.* It flowers during June. It is one of the most offensive hay fever grasses.

2. Anthoxanthum aristatum Bioss. Voy. Bot. Esp. 2:638. 1845. *Fig. 196.*

Anthoxanthum puelii Lec. & Lam. Cat. Pl. France 385. 1847. Annual; culms decumbent, to 35 cm tall; blades more or less scabrous, 2–5 mm broad; panicle to 4 cm long, exserted; spikelets whitish-green, 5–7 mm long; glumes glabrous, the first short-awned; fertile lemma awnless, 1.0–1.5 mm long; awns of sterile lemmas long-exserted; 2n = 10 (Avdulov, 1928).

195. *Anthoxanthum odoratum* (Sweet Vernal Grass). *a.* Inflorescences, X½. *b.* Sheath, with ligule, X2½. *c.* Spikelet, X5. *d.* First glume, X7½. *e.* Second glume, X7½. *f.* Lemmas, X7½.

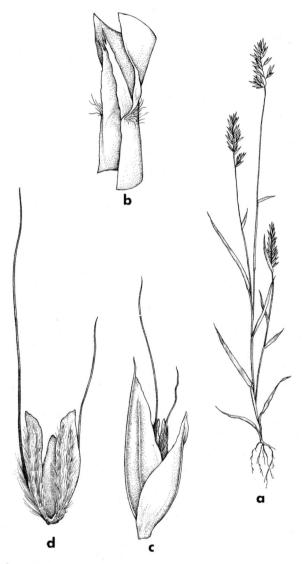

196. Anthoxanthum aristatum (Annual Sweet Grass). *a.* Habit, X½. *b.* Sheath, with ligule, X5. *c.* Spikelet, X7½. *d.* Lemmas, X7½.

COMMON NAME: Annual Sweet Grass.

HABITAT: Waste ground.

RANGE: Native of Europe; sparingly introduced in the United States.

ILLINOIS DISTRIBUTION: Rare; known as an adventive in Rock Island County (along railroad, Rock Island, August 18, 1963, *R. H. Mohlenbrock 13134*).

21. Hierochloë R. BR. – Sweet Grass

Rhizomatous perennials; blades flat; inflorescence paniculate; spikelets with 1 perfect flower and 2 staminate flowers below the perfect ones, disarticulating above the glumes; glumes equal, papery, 3-nerved; lemmas more or less indurate, rounded on the back, awnless.

The staminate lower florets distinguish this genus from *Anthoxanthum* and *Phalaris* in Illinois.

Only the following species occurs in Illinois.

1. Hierochloë odorata (L.) Beauv. Ess. Agrost. 62. 1812. *Fig. 197.*

Holcus odoratus L. Sp. Pl. 1048. 1753.

Holcus fragrans Willd. Sp. Pl. 4:936. 1806.

Holcus borealis Schrad. Fl. Germ. 1:252. 1806.

Hierochloë borealis (Schrad.) Roem. & Schult. Syst. Veg. 2:513. 1817.

Hierochloë fragrans (Willd.) Roem. & Schult. Syst. Veg. 2:514. 1817.

Torresia odorata (L.) Hitchc. Am. Journ. Bot. 2:301. 1915.

Perennial from slender, creeping rhizomes; culms to 60 cm tall; sheaths often bladeless; cauline leaves 2–3, scabrous, 2–5 mm broad; panicle pyramidal, to 10 cm long, the branches spreading, or the lowermost drooping; spikelets 4.5–6.0 mm long; glumes broadly ovate, glabrous, 4–6 mm long; fertile lemma 3–5 mm long, pubescent near apex; staminate lemmas 4–6 mm long, acute, pubescent on margin and near summit of keel; $2n = 28, 56$ (Myers, 1947).

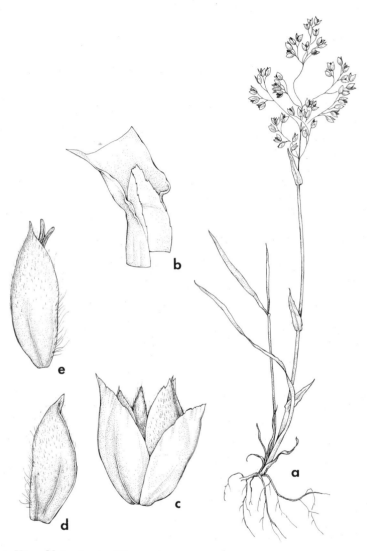

197. Hierochloë odorata (Sweet Grass). *a.* Habit, X½. *b.* Sheath, with ligule, X4. *c.* Spikelet, X7½. *d.* Fertile lemma, X7½. *e.* Staminate lemma, X7½.

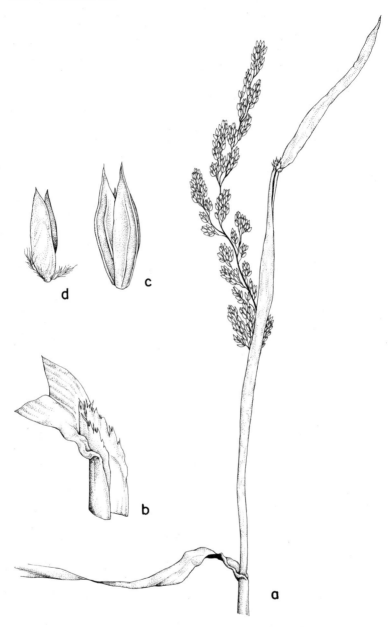

198. Phalaris arundinacea (Reed Canary Grass). *a.* Inflorescence, X½. *b.* Sheath, with ligule, X2½. *c.* Spikelet, X6½. *d.* Lemmas, X6½.

COMMON NAME: Sweet Grass.

HABITAT: Meadows, usually in moist situations.

RANGE: Greenland to Alaska, south to Arizona, Illinois, and New Jersey; Europe; Asia.

ILLINOIS DISTRIBUTION: Occasional in the northern one-fifth of the state; absent elsewhere.

This predominantly northern grass derives its common name from its sweet fragrance. It flowers during May and June.

22. *Phalaris* L. – Canary Grass

Annuals or perennials; blades flat; panicles contracted, spike-like; spikelets with 1 terminal perfect flower and 1–2 empty lemmas below, disarticulating above the glumes; glumes subequal, keeled; sterile lemmas minute, awnless; fertile lemma indurate (at least in fruit), awnless; palea smaller than the lemma, 2-nerved.

KEY TO THE SPECIES OF Phalaris IN ILLINOIS

1. Keel of glumes wingless, the glumes 4.5–6.5 mm long; sterile florets 1–2 mm long; perennial from creeping rhizomes＿＿1. *P. arundinacea*
1. Keel of glumes broadly winged, the glumes 7–10 mm long; sterile florets 2.5–4.5 mm long; annual＿＿＿＿＿＿＿＿＿＿＿＿＿＿2. *P. canariensis*

1. Phalaris arundinacea L. Sp. Pl. 55. 1753. *Fig. 198.*

Phalaris arundinacea var. *picta* L. Sp. Pl. 55. 1753.

Phalaris arundinacea var. *variegata* Parnell, Grasses Brit. 188. 1845.

Phalaris arundinacea f. *variegata* (Parnell) Druce, Fl. Berks. 556. 1897.

Phalaris arundinacea f. *picta* (L.) Asch. & Graebn. Syn. Mitt. Eur. Fl. 2:24. 1898.

Phalaris arundinacea ssp. *typica* Ruiz, Anal. Jard. Bot. Madrid. 8:489. 1947.

Perennial from scaly, creeping rhizomes; culms to 1.5 m tall; blades 10–20 mm broad, green or occasionally white-striped (f. *picta*); panicle contracted, to 30 cm long, lobed at base; glumes 4.5–6.5 mm long, acute, wingless, the keel scabrous; fertile floret 2.7–4.5 mm long, lanceolate, glabrous or appressed-pubescent, conspicuously nerved; sterile florets 2, subulate, pubescent, 1–2 mm long; 2n = 14, 28 (Church, 1929).

COMMON NAME: Reed Canary Grass.

HABITAT: Meadows and similar moist situations.

RANGE: Newfoundland to Alaska, south to California, Colorado, Illinois, and North Carolina.

ILLINOIS DISTRIBUTION: Occasional throughout the state. Ribbon Grass (f. *picta*), the cultivated form with white-striped leaves, occasionally escapes around old dwellings. It is known from Cook, Ford, Lake, and St. Clair counties. *Phalaris arundinacea* varies considerably in leaf width. The species flowers from mid-May to mid-July.

2. Phalaris canariensis L. Sp. Pl. 54. 1753. *Fig. 199.*

Annual; culms erect, to 1 m tall; blades to 15 mm broad; panicle contracted, ovoid, to 4 cm long; spikelets 5–6 mm long; glumes 7–10 mm long, striate, the keel broadly winged, strigose or glabrous; fertile floret acute, 4.8–6.8 mm long, densely appressed-pubescent; sterile florets 2, 2.5–4.5 mm long, sparsely pubescent; 2n = 12 (Avdulov, 1928).

COMMON NAME: Canary Grass.

HABITAT: Waste ground.

RANGE: Native of Europe; introduced throughout the United States.

ILLINOIS DISTRIBUTION: Not common; sparse throughout the state. This species flowers during June and July. It differs from *P. arundinacea* in its annual habit, its broadly winged keel of the glumes, and its larger sterile lemmas.

23. Alopecurus L. – Foxtail

Annuals or perennials; blades flat; inflorescence a dense panicle, appearing spike-like; spikelets 1-flowered, disarticulating below the glumes; glumes equal, keeled; lemma 5-nerved, awned, the margins partly connate; palea none.

KEY TO THE SPECIES OF Alopecurus IN ILLINOIS

1. Spike-like panicle 6–10 mm thick; spikelets 4.0–5.5 mm long; glumes acute, 4.0–5.5 mm long; lemma 4.0–5.5 mm long, the awn exserted 4–5 mm_____1. *A. pratensis*
1. Spike-like panicle to 5.5 mm thick; spikelets 2.0–2.7 mm long;

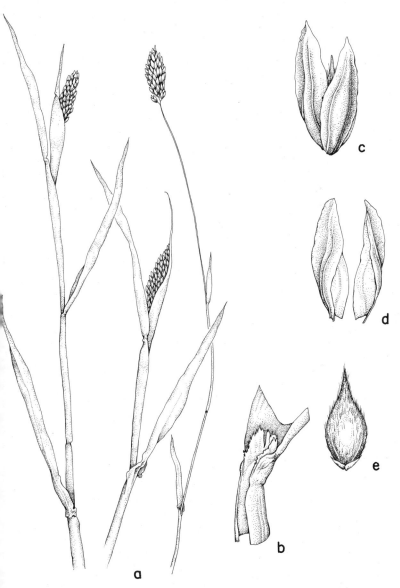

199. Phalaris canariensis (Canary Grass). *a.* Inflorescences, X½. *b.* Sheath, with ligule, X2½. *c.* Spikelet, X5. *d.* Glumes, X5. *e.* Fertile floret, X5.

glumes obtuse, 2.0–2.7 mm long; lemma 2.0–2.7 mm long, the awn exserted only up to 4 mm.

2. Perennial from slender rhizomes; awn exserted up to 1 mm____
_____2. A. *aequalis*
2. Tufted annual; awn exserted from 1.5–4.0 mm_____
_____3. A. *carolinianus*

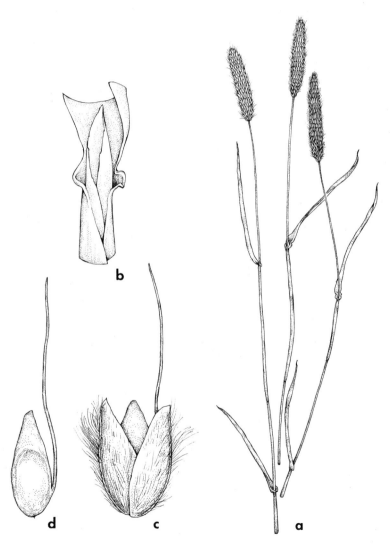

200. *Alopecurus pratensis* (Meadow Foxtail). *a.* Inflorescences, X½. *b.* Sheath, with ligule, X4. *c.* Spikelet, X6. *d.* Lemma, X6.

1. Alopecurus pratensis L. Sp. Pl. 60. 1753. *Fig. 200.*

Perennial from creeping rhizomes; culms erect or decumbent, glabrous, to 80 cm long, 6–10 mm thick; spikelets 4.0–5.5 mm long; glumes acute, 4.0–5.5 mm long, ciliate on the narrowly winged keel; lemma 4.5–5.5 mm long, subacute, 5-nerved, the awn exserted 4–5 mm; 2n = 28 (Marchal, 1920), 42 (Johnsson, 1941).

COMMON NAME: Meadow Foxtail.
HABITAT: Along railroads (in Illinois); mowed fields.
RANGE: Native of Eurasia; adventive in the northern United States.
ILLINOIS DISTRIBUTION: Rare; known from four counties in the northern half of the state. A specimen from Peoria County was collected as early as 1875.
The Illinois collections were made in June. Meadow Foxtail begins to flower a few weeks before Timothy, a species which it strongly resembles.

2. Alopecurus aequalis Sobol. Fl. Petrop. 16. 1799. *Fig. 201.*

Alopecurus aristulatus Michx. Fl. Bor. Am. 1:43. 1803.
Alopecurus geniculatus var. *aristulatus* (Michx.) Torr. Fl. N. & Mid. U. S. 1:97. 1824.

Tufted perennial; culms erect or decumbent, glabrous, to 50 cm long; blades 1–4 mm broad; spike-like panicle to 7.5 cm long, 3.0–5.5 mm thick; spikelets 2.0–2.7 mm long; glumes obtuse, 2.0–2.7 mm long, villous at base, ciliate on keel; lemma 2.0–2.7 mm long, obtuse, 5-nerved, the awn exserted up to 1 mm; 2n = 14 (Avdulov, 1931).

COMMON NAME: Foxtail.
HABITAT: Wet ground; occasionally in shallow water.
RANGE: Newfoundland to Alaska, south to California, New Mexico, Kansas, and Maryland; Europe; Asia.
ILLINOIS DISTRIBUTION: Not common; scattered throughout the state.
This small and perhaps occasionally overlooked species flowers from May to July. The short awns and the perennial habit distinguish it from *A. carolinianus*.
Fernald (1930) has discussed the identity of this species. Early Illinois workers knew this species as *A. aristulatus* or *A. geniculatus* var. *aristulatus*.

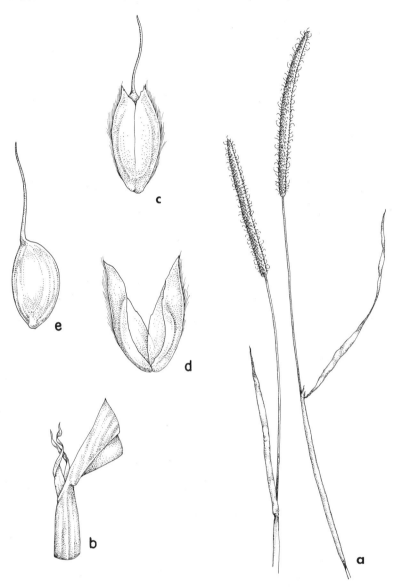

201. *Alopecurus aequalis* (Foxtail). *a.* Inflorescences, X½. *b.* Sheath, with ligule, X4. *c.* Spikelet, X10. *d.* Glumes, X11. *e.* Lemma, X10.

3. Alopecurus carolinianus Walt. Fl. Carol. 74. 1788. *Fig. 202.*

Alopecurus ramosus Poir. in Lam. Encycl. 8:776. 1808.
Alopecurus geniculatus var. *ramosus* (Poir.) St. John, Rhodora 19:167. 1917.

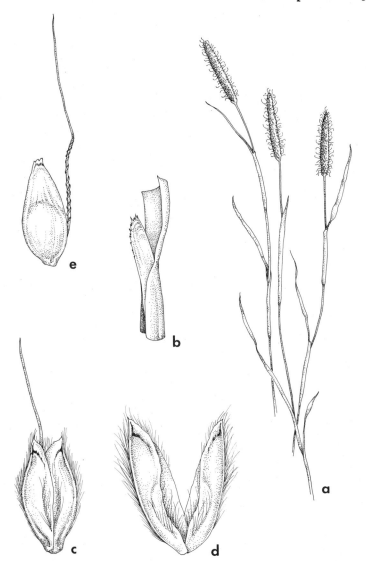

202. *Alopecurus carolinianus* (Common Foxtail). *a.* Inflorescences, X½. *b.* Sheath, with ligule, X4. *c.* Spikelet, X12½. *d.* Glumes, X15. *e.* Lemma, X15.

Tufted annual; culms usually erect, glabrous, to 60 cm tall; blades 1–4 mm broad; spike-like panicle to 5 cm long, 4–5 mm thick; spikelets 2.0–2.7 mm long; glumes obtuse, 2.0–2.6 mm long, villous at base, ciliate on keel; lemma 2.0–2.7 mm long, obtuse, 5-nerved, the awn exserted 1.5–4.0 mm.

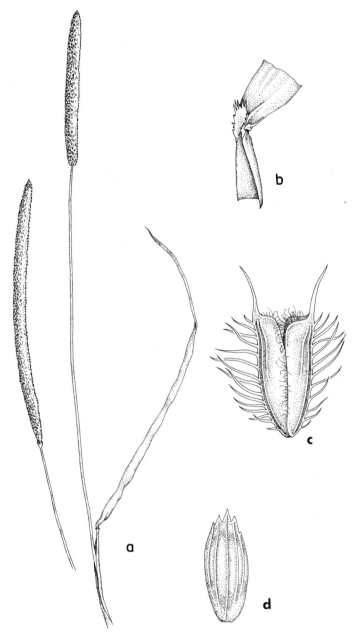

203. *Phleum pratense* (Timothy). *a*. Inflorescences, X½.   *b*. Sheath, with ligule, X2½.   *c*. Spikelet, X12½.   *d*. Lemma, X12½.

COMMON NAME: Common Foxtail.

HABITAT: Moist ground.

RANGE: New York to British Columbia, south to California, Texas and Florida.

ILLINOIS DISTRIBUTION: Occasional throughout the state. The Common Foxtail flowers from late April to early July.

The longer awns of the glumes distinguish this species from *A. aequalis,* while the shorter spikelets distinguish it from *A. pratensis.*

24. *Phleum* L. – Timothy

Perennials; blades flat; panicle dense, spike-like; spikelets 1-flowered, disarticulating above the glumes; glumes equal, keeled, awned; lemma smaller than the glumes, 3- to 5-nerved, awnless; palea nearly as long as the lemma.

Only the following species occurs in Illinois.

1. **Phleum pratense** L. Sp. Pl. 59. 1753. *Fig. 203.*

Cespitose perennial; culms erect, simple, glabrous except for the scabrous apex, to 85 cm tall; sheaths glabrous; blades 4–8 mm broad, scabrous along the margins; spike-like panicle cylindric, to 10 (–15) cm long, 6–8 (–10) mm thick; spikelets 2.5–3.5 mm long, crowded; glumes 2.5–3.5 mm long, rounded or truncate at the apex, ciliate on the keel, the awn to 1.5 mm long; $2n = 42$ (Myers, 1944).

COMMON NAME: Timothy.

HABITAT: Waste ground, fields.

RANGE: Native of Europe; frequently cultivated and often escaped in the United States.

ILLINOIS DISTRIBUTION: In every county.

Timothy is the most important hay grass in the United States.

The flowers, which are produced from early June to mid-August, form abundant pollen which causes much hay fever in the area.

25. *Milium* L. – Millet Grass

Tall perennials; blades flat; panicle open; spikelets 1-flowered, disarticulating above the glumes; glumes equal, rounded on the back, 3-nerved; lemmas indurate, more or less compressed, essentially nerveless, the margins partly inrolled around the palea, indurate, nerveless.

The indurate lemma and palea give this genus a striking resemblance to *Panicum*. The absence of awns on the glumes and lemmas distinguishes *Milium* from the other genera with indurate lemmas.

Only the following species occurs in Illinois.

1. Milium effusum L. Sp. Pl. 61. 1753. *Fig. 204.*

Perennial; culms erect, bent at base, glabrous, often glaucous, to 1.5 m tall; blades flat, to 20 mm broad, glabrous or scabrous; panicle ovoid, to 25 cm long, the branches spreading or somewhat reflexed; spikelets 3.0–3.5 mm long; glumes ovate to elliptic, obtuse to acute, scaberulous, 2.5–3.5 mm long; palea indurate; 2n = 28 (Avdulov, 1928).

COMMON NAME: Millet Grass.
HABITAT: Moist woodlands.
RANGE: Newfoundland to Minnesota, south to Illinois and Maryland; Eurasia.
ILLINOIS DISTRIBUTION: Rare; known from Kane (*Vasey s.n.*) and Tazewell (*Brendel s.n.*) counties, and not seen since the nineteenth century in Illinois.
This species, which may be extinct in Illinois, flowers from June to August.

26. *Beckmannia* HOST – Slough Grass

Stout annuals; blades flat; inflorescence paniculate, composed of ascending spikes; spikelets 1-flowered (in Illinois), compressed, disarticulating below the glumes; glumes subequal, papery, inflated; lemma firmer but narrower than the glumes, 5-nerved, partly enclosing the slightly shorter, rigid palea.

Only the following species occurs in Illinois. A discussion of this species has been presented by Fernald (1928).

1. Beckmannia syzigachne (Steud.) Fern. Rhodora 30:27. 1928. *Fig. 205.*

Panicum syzigachne Steud. Flora 29:19. 1846.
Annuals; culms solitary or tufted, to nearly 1 m tall; blades flat, light green, scaberulous, to 8 mm broad; panicle slender, erect, to 25 cm long, composed of strongly ascending spikes to 1 cm long; spikelets 2–3 mm long, equally as broad, with 1 perfect flower and sometimes one imperfect flower; glumes broadly triangular, wrinkled, keeled, cuspidate, 3-nerved, 2–3 mm long, the

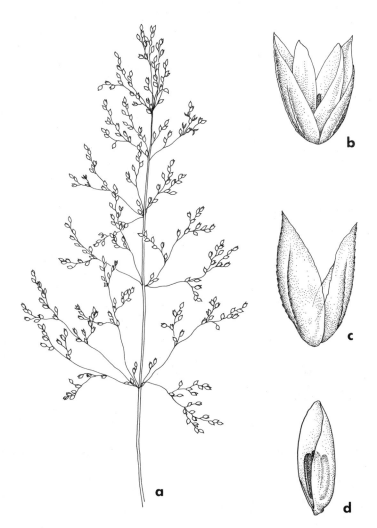

204. Milium effusum (Millet Grass). *a.* Inflorescence, X½. *b.* Spikelet, X10. *c.* Glumes, X10. *d.* Lemma, X10.

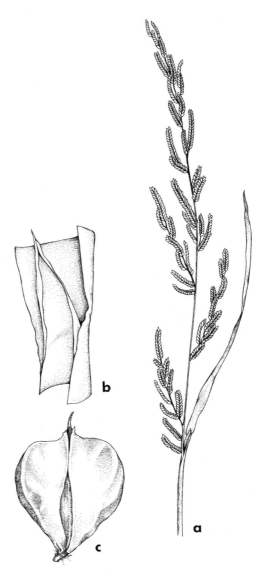

205. *Beckmannia syzigachne* (American Slough Grass). *a*. Inflorescence, X½. *b*. Sheath, with ligule, X5. *c*. Spikelet, X12½.

margins sometimes or nearly meeting; lemma lanceolate, acuminate, 2–3 mm long, 5-nerved; 2n = 14 (Avdulov, 1931).

COMMON NAME: American Slough Grass.

HABITAT: Wet ground.

RANGE: Quebec to Alaska, south to California, New Mexico, and Illinois; probably adventive in New York and Pennsylvania; Asia.

ILLINOIS DISTRIBUTION: Very rare; first collected from Clyde, Cook County, by Umbach during the nineteenth century, and not seen again in Illinois until it was discovered by Floyd Swink in 1955 (Lake Co.: marshy ground on the south shore of Loon Lake near the village of Loon Lake, July 23, 1955, *Swink 2772*).

The very broad spikelets readily distinguish this species which flowers during the summer.

Tribe *Triticeae*

Mostly perennials; inflorescence a spike or spike-like raceme; spikelets (1-) 2- to several-flowered; lemmas 5- to 7-nerved, awned or awnless.

This small tribe is of great economic importance because it contains most of the cereal grasses. Seven genera comprise this tribe in Illinois.

27. *Elymus* L. – Wild Rye

Tall perennials; blades flat or involute; inflorescence spicate, densely or loosely flowered, erect or nodding; spikelets 1- to 6-flowered, 1–4 at each joint of the rachis, disarticulating above the glumes; glumes 2 (or absent), rigid, 1- to 5-nerved, sometimes awned; lemmas indurate, rounded on the back, usually 5-nerved, usually awned.

Various detailed treatments of *Elymus* may be found by Wiegand (1918), Fernald (1933), and Church (1967, 1967a). Several of the species are highly variable, while others tend to hybridize.

Hystrix is included within *Elymus* in this treatment on the evidence presented by Church (1967, 1967a).

KEY TO THE TAXA OF Elymus IN ILLINOIS
(MODIFIED FROM CHURCH [1967])

1. Rhizomes present; glumes 3–4 mm broad; lemmas awnless_____
_____1. E. arenarius
1. Rhizomes absent; glumes up to 2.5 mm broad; lemmas awned (awnless in one variety of E. virginicus).
   2. Glumes reduced to unequal filiform bristles, or absent; spikelets widely spreading_____2. E. hystrix
   2. Glumes subequal in length; spikelets ascending.
      3. Base of glumes indurated; paleas up to 8.5 mm long.
         4. Glumes (0.8–) 1.0–2.5 mm wide, swollen on at least half or all of the adaxial surface_____3. E. virginicus
         4. Glumes 0.2–1.0 mm wide, indurated for 1–3 mm adaxially.
            5. Paleas 7.0–8.5 mm long, the apices bidentate_____
_____4. E. riparius
            5. Paleas 5.5–6.5 mm long, the apices obtuse_____
_____5. E. villosus
      3. Base of glumes thin or indurated for only 1 mm; paleas 8.5–12.0 (–14.0) mm long_____6. E. canadensis

1. Elymus arenarius L. Sp. Pl. 83. 1753. *Fig. 206.*

Perennial from a short rhizome; culms to 1.2 m tall, glabrous above, arising from several old leaf bases; blades firm, flat or involute, glaucous, 5–15 mm broad, scaberulous above; spikes stiff, dense, erect, 7–25 cm long, 1–3 cm broad; spikelets borne in pairs, 20–30 mm long, 3- to 7-flowered; glumes lanceolate, acuminate or mucronate, pilose or scabrous on the keel, glabrous at the base, (1-) 3- to 5-nerved, 15–35 mm long, 3–4 mm broad; lemmas awnless, acuminate or mucronate, villous or scabrous, 15–30 mm long.

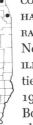

COMMON NAME: Wild Rye; Lyme Grass.
HABITAT: Sandy shores of Lake Michigan.
RANGE: Europe; Asia; introduced in Greenland; Canada; New York, Illinois, Wisconsin.
ILLINOIS DISTRIBUTION: Rare; only Cook and Lake counties; originally found in Illinois in 1916 and refound in 1952.
Bowden (1957) conclusively shows that the Illinois plants represent the introduced *Elymus arenarius* and not the much more northern *Elymus mollis* Trin. in Spreng. *Elymus mollis* differs by having the glumes pubescent throughout.

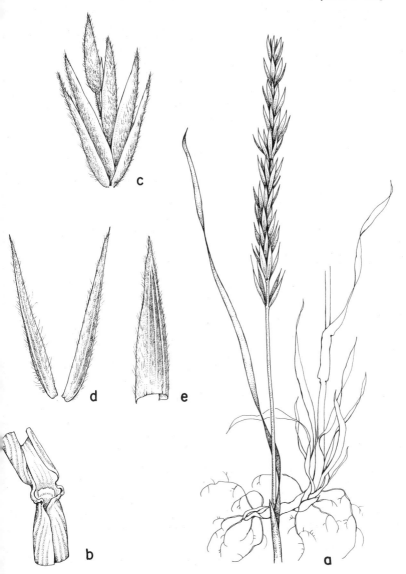

206. *Elymus arenarius* (Wild Rye). *a.* Habit, X½. *b.* Sheath, with ligule, X2½. *c.* Spikelet, X2. *d.* Glumes, X3. *e.* Lemma, X3.

This is the only Illinois species of *Elymus* with rhizomes and,
except for *E. virginicus* var. *submuticus,* the only taxon with awnless lemmas.

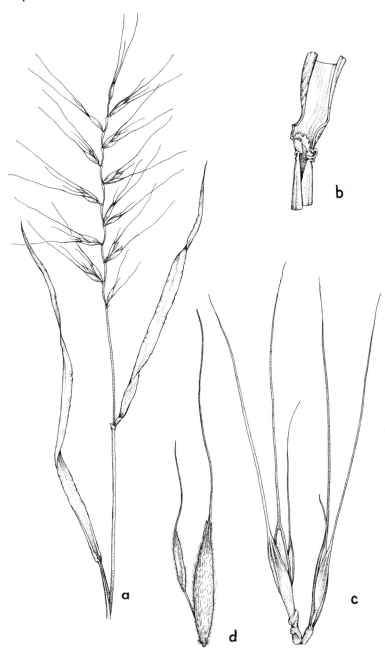

207. *Elymus hystrix* (Bottlebrush Grass).—var. *hystrix*. a. Inflorescence, X1. b. Sheath, with ligule, X2½. c. Spikelets, X3.—var. *bigeloviana*. d. Lemmas, X3.

2. Elymus hystrix L. Sp. Pl. 560. 1753.

Hystrix patula Moench, Meth. Pl. 295. 1794.

Asperella hystrix (L.) Willd. Enum. Pl. 132. 1809.

Gymnostichum hystrix (L.) Schreb. Beschr. Gras. 2:127. 1810.

Hystrix hystrix (L.) Millsp. Fl. W. Va. 474. 1892.

Culms solitary or few, to 1.2 m tall; sheaths glabrous or scabrous or pubescent; blades glabrous or scabrous or pubescent, 7–15 mm broad; spikes 5–20 cm long, erect or slightly arching; spikelets usually remote, usually paired, 2- to 4-flowered, 25–55 mm long (including the awns); glumes up to 15 mm long, setiform, or absent from the uppermost spikelets; lemmas 18–50 mm long (including the curved, rough awns), glabrous or pubescent; 2n = 28 (Brown, 1948).

COMMON NAME: Bottlebrush Grass.

HABITAT: Woodlands.

Two variations and two natural hybrids for which *E. hystrix* is one of the parents occur in Illinois. These taxa may be separated as follows:

1. Paleas without cilia; glumes lacking or usually not exceeding 15 mm in length.
   2. Lemmas glabrous_____2a. *E. hystrix* var. *hystrix*
   2. Lemmas pubescent_____2b. *E. hystrix* var. *bigeloviana*
1. Paleas ciliate; glumes present, most of them usually longer than 15 mm_____(See hybrid taxa after discussion of 2b.)

2a. Elymus hystrix L. var. **hystrix** *Fig. 207a–c.*

Paleas without cilia; glumes lacking or usually not exceeding 15 mm in length; lemmas glabrous.

RANGE: Maine to North Dakota, south to Oklahoma and Georgia.

ILLINOIS DISTRIBUTION: Common; in every county.

This attractive grass flowers from June to mid-August. This variety, with glabrous lemmas, is the more widespread. It differs from the hybrids it forms with *E. virginicus* by its eciliate paleas.

A collection by Brendel from Peoria County, on which Mosher (1918) bases her report of *E. diversiglumis* Scribn. & Ball, is a more unusual morphological variant of *E. hystrix* in which long, filiform glumes are found all along the more or less sinuous rachis of rather closely compacted nodes. Church (1967) cites two other Illinois collections of this kind: *Chase 1840* from Knox County; *Chase 12053* from Peoria County.

2b. Elymus hystrix L. var. **bigeloviana** (Fern.) Mohlenbrock, comb. nov. *Fig. 207d.*

Asperella hystrix var. *bigeloviana* Fern. Rhodora 24:230. 1922.
Hystrix patula Moench var. *bigeloviana* (Fern.) Deam, Pub. Ind. Dept. Conserv. 82:117. 1929.

Paleas without cilia; glumes lacking or usually not exceeding 15 mm in length; lemmas pubescent.

RANGE: Nova Scotia to Wisconsin, south to Illinois and Pennsylvania.

ILLINOIS DISTRIBUTION: Occasional throughout the state (not mapped).

Two different natural hybrids have been collected from the Pine Hills of Union County in which *Elymus hystrix* var. *hystrix* is one of the parents. These hybrids have been thoroughly discussed by Church (1967a) who has the experimental evidence to prove the parentage of the hybrids. Both hybrids are the results of crosses between *E. hystrix* var. *hystrix* and *E. virginicus* var. *glabriflorus*. The hybrids are intermediate between the two parents in that they possess relatively short, filiform glumes, rather loosely disposed spikelets, and ciliated paleas. Both forms of *E. virginicus* var. *glabriflorus* known from Illinois hybridize with *E. hystrix*. When *E. virginicus* var. *glabriflorus* f. *glabriflorus* hybridizes with *E. hystrix*, the product has spikelets with glabrous lemmas (*Fig. 208*). When *E. virginicus* var. *glabriflorus* f. *australis* hybridizes with *E. hystrix*, the lemmas of the offspring are hirsute (*Fig. 209*). Specimens collected in the past from southern Illinois which have as their parents *E. virginicus* var. *glabriflorus* f. *glabriflorus* and *E. hystrix* have been confused with *E. interruptus*, a taxon apparently not present in the Illinois flora.

3. Elymus virginicus L. Sp. Pl. 84. 1753.

Densely cespitose perennial to 1.5 m tall; sheaths glabrous (rarely pubescent); blades scabrous, green or glaucous, 3–15 mm broad; spikes stiff, dense, erect, 5–17 cm long, 1–3 cm broad; spikelets 2- to 4-flowered; glumes indurate, flat, scabrous, 4- to 5-nerved, acuminate or short-awned, 10–40 mm long, 1.0–2.5 mm broad; lemmas glabrous to villous, nerveless below, nerved above, 10–45 mm long (including the straight awns).

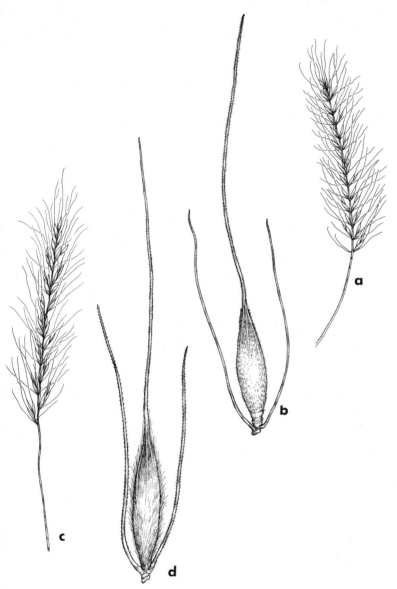

208. Hybrid Taxa.—*a, b. Elymus hystrix* × *E. virginicus* var. *glabriflorus*
f. *glabriflorus.—c, d. Elymus hystrix* × *E. virginicus* var. *glabriflorus* f. *australis.*

COMMON NAME: Virginia Wild Rye; Lyme Grass.
HABITAT: Fields and woodlands; low ground.
This is a common and highly variable species. Many of the variations are more clear-cut than those of *E. canadensis;* therefore, recognition of certain of these variations is made in this work. All variations occur here and there throughout the state, and no effort has been made to map the different taxa. For a discussion of the hybrids formed between this species and *E. hystrix,* see species 2.

KEY TO THE TAXA OF Elymus virginicus IN ILLINOIS

1. Lemmas 10–30 mm long; glumes 10–25 (–27) mm long.
  2. Glumes and lemmas awned.
    3. Glumes and lemmas glabrous or scabrous_____
    _____3a. *E. virginicus* var. *virginicus* f. *virginicus*
    3. Glumes and lemmas villous_____
    _____3b. *E. virginicus* var. *virginicus* f. *hirsutiglumis*
  2. Glumes and lemmas acuminate or subulate-tipped_____
    _____3c. *E. virginicus* var. *submuticus*
1. Lemmas (30–) 35–45 mm long; glumes 27–40 mm long.
  4. Glumes and lemmas glabrous or scabrous_____
    _____3d. *E. virginicus* var. *glabriflorus* f. *glabriflorus*
  4. Glumes and lemmas hirsute_____
    _____3e. *E. virginicus* var. *glabriflorus* f. *australis*

3a. Elymus virginicus L. var. **virginicus** f. **virginicus** *Fig. 209a–d.*

Elymus virginicus f. *jejunus* Ramaley, Minn. Bot. Stud. 9:114. 1894.
Elymus jejunus (Ramaley) Rydb. Bull. Torrey Club 36:539. 1909.
Elymus virginicus var. *jejunus* (Ramaley) Bush, Am. Midl. Nat. 10:65. 1926.
Elymus virginicus var. *typicus* Fern. Rhodora 35:198. 1933.
Glumes 10–25 (–27) mm long, awned, glabrous or scabrous; lemmas 10–30 mm long, awned, glabrous or scabrous; 2n = 28 (Brown, 1948).

RANGE: Newfoundland to Alberta, south to Texas and North Carolina.

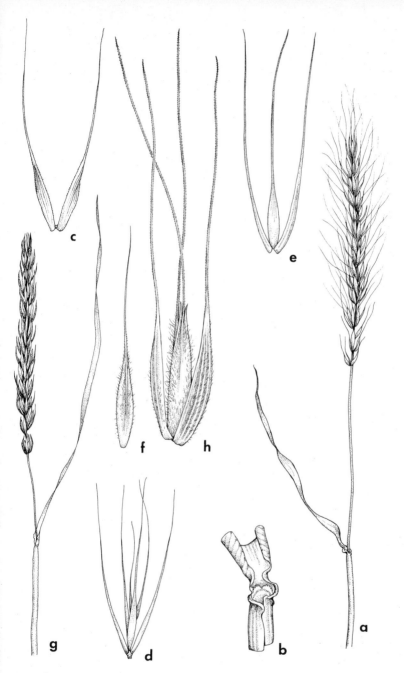

209. *Elymus virginicus* (Virginia Wild Rye).—var. *virginicus.* *a.* Inflorescence, X½. *b.* Sheath, with ligule, X2½. *c.* Glumes, X2. *d.* Spikelet, X1½.—var. *glabriflorus.* *e.* Spikelet, X2.—forma *hirsutiglumis.* *f.* Lemma, X2.—var. *submuticus.* *g.* Inflorescence, X½.—forma *australis.* *h.* Spikelet, X3½.

This includes var. *jejunus*, a taxon in which the spikes are well-exserted from the sheaths. In var. *virginicus*, the spikes are said to be included or barely exserted from the sheaths. There appears to be too much overlapping to justify recognition of var. *jejunus*.

3b. **Elymus virginicus** L. var. **virginicus** f. **hirsutiglumis** (Scribn.) Fern. Rhodora 35:198. 1933. *Fig. 209f.*

Elymus canadensis var. *intermedius* Vasey ex Gray, Man., ed. 6, 673. 1890.

Elymus intermedius (Vasey) Scribn. & Smith, Bull. U.S.D.A. Div. Agrost. 4:38. 1897.

Elymus hirsutiglumis Scribn. Bull. U.S.D.A. Div. Agrost. 11:58. 1898.

Elymus virginicus var. *hirsutiglumis* (Scribn.) Hitchcock, Rhodora 10:65. 1908.

Similar to the preceding form, except that the glumes and lemmas are villous in f. *hirsutiglumis*.

> RANGE: Quebec to North Dakota, south to Texas and Virginia.

3c. **Elymus virginicus** L. var. **submuticus** Hook. Fl. Bor. Amer. 2:255. 1840. *Fig. 209g.*

Elymus submuticus (Hook.) Smyth, Trans. Kans. Acad. Sci. 25:99. 1913.

Elymus virginicus f. *submuticus* (Hook.) Pohl, Amer. Midl. Nat. 38:549. 1947.

Similar to var. *virginicus*, except that the glumes and lemmas are merely acuminate or subulate-tipped in var. *submuticus*.

> RANGE: Quebec to Washington, south to Oklahoma, Illinois, and Rhode Island.

This is the least common variation in Illinois.

3d. **Elymus virginicus** L. var. **glabriflorus** (Vasey) Bush f. **glabriflorus** *Fig. 209e.*

Elymus canadensis var. *glabriflorus* Vasey ex L. H. Dewey, Contrib. U. S. Nat. Herb. 2:550. 1894.

Elymus glabriflorus (Vasey) Scribn. & Ball, Bull. U.S.D.A. Div. Agrost. 24:49. 1901.

Elymus australis var. *glabriflorus* (Vasey) Wiegand, Rhodora 20:84. 1918.

Elymus virginicus L. var. *glabriflorus* (Vasey) Bush, Am. Midl. Nat. 10:62. 1926.

Glumes 27–40 mm long, awned, glabrous or scabrous; lemmas (30–) 35–45 mm long, awned, glabrous or scabrous; 2n = 28 (Brown, 1948).

RANGE: Maine to Nebraska, south to New Mexico and Florida.

This form hybridizes naturally with *E. hystrix.*

3e. **Elymus virginicus** L. var. **glabriflorus** (Vasey) Bush f. **australis** (Scribn. & Ball) Fern. Rhodora 35:198. 1933. *Fig. 209h.*

Elymus australis Scribn. & Ball, U.S.D.A. Div. Agrost. 24:46. 1901.

Elymus virginicus var. *australis* (Scribn. & Ball) Hitchcock in Deam, Pub. Ind. Dept. Conserv. 82:113. 1929.

Similar to f. *glabriflorus,* except that the glumes and lemmas are hirsute in f. *australis.*

RANGE: As in var. *glabriflorus.*

This form hybridizes naturally with *E. hystrix.*

4. **Elymus riparius** Wiegand, Rhodora 20:84. 1918. *Fig. 210.*

Cespitose perennial with rather slender culms to 1.5 m tall; sheaths glabrous or scaberulous; blades glabrous or scaberulous, thin, green or glaucous, 5–20 mm broad; spikes slightly nodding, 6–20 cm long, 2–4 cm broad; spikelets 2- to 4-flowered; glumes setiform, indurate, 3-nerved, 17–30 mm long (including the awns), less than 1 mm broad; lemmas hispidulous to nearly glabrous, 22–45 mm long (including the straight awn); palea 7.5–8.0 mm long; 2n = 28 (Brown, 1948).

COMMON NAME: Wild Rye.

HABITAT: Woodlands.

RANGE: Quebec to Wisconsin, south to Arkansas and Florida.

ILLINOIS DISTRIBUTION: Not common; scattered in several counties.

This species has the glabrous or scabrous sheaths and blades of *E. virginicus* but the nodding spike and setiform glumes of *E. villosus.*

It flowers from early July to mid-September.

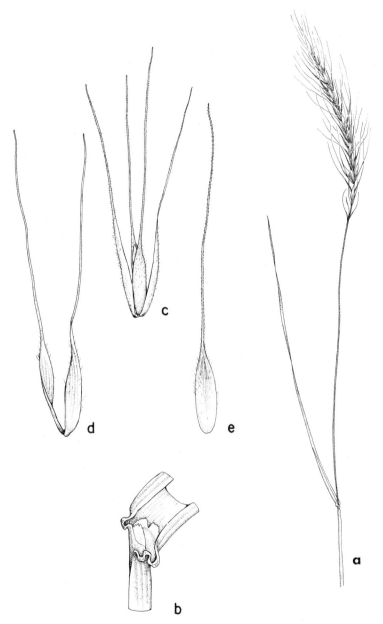

210. *Elymus riparius* (Wild Rye). *a.* Inflorescence, X½. *b.* Sheath, with ligule, X2½. *c.* Spikelet, X2. *d.* Glumes, X2. *e.* Lemma, X2½.

An excellent discussion of this species has been presented by Church (1954).

The first collection from Illinois was made by Wolf from Fulton County during the middle of the nineteenth century.

5. Elymus villosus Muhl. in Willd. Enum. Pl. 1:131. 1809.

Perennial in small tufts with culms to 1 m tall; sheaths villous, rarely glabrous; blades thin, flat, usually villous above, glabrous or scaberulous beneath, 5–10 mm broad; spikes dense, nodding, 5–15 cm long, 2–3 cm broad; spikelets 1- (2-) flowered; glumes setiform, 1- to 3-nerved, hispid or hirsute or glabrous, 15–30 mm long (including the awns), less than 1 mm broad; lemmas villous or glabrous or scabrous, 25–45 mm long (including the straight awn); palea 5.0–6.5 mm long.

COMMON NAME: Slender Wild Rye.

HABITAT: Woodlands.

Two variations may be recognized in Illinois.

1. Glumes and lemmas hispid to hirsute_____5a. *E. villosus* f. *villosus*
1. Glumes scabrous; lemmas glabrous or scabrous_____
_____5b. *E. villosus* f. *arkansanus*

5a. Elymus villosus Muhl. f. villosus *Fig. 211a–e.*

Elymus striatus var. *villosus* (Muhl.) Gray, Man. 603. 1848.
Elymus propinquus Fresen. ex Steud. Syn. Pl. Glum. 1:349. 1854.

Glumes and lemmas hispid to hirsute.

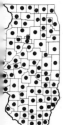

RANGE: Quebec to Wyoming, south to Texas and Georgia.

ILLINOIS DISTRIBUTION: Common; in every county.

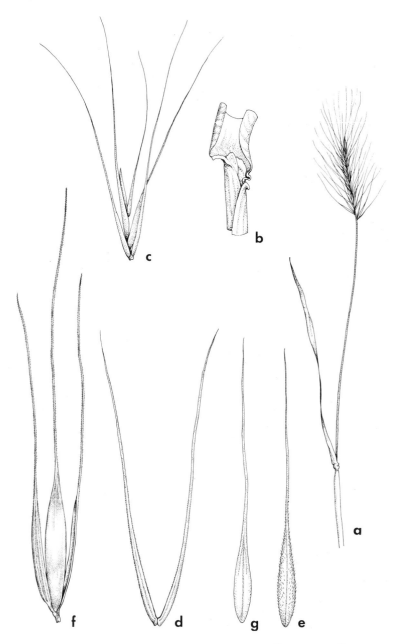

211. *Elymus villosus* (Slender Wild Rye).—var. *villosus.* *a.* Inflorescence, X½. *b.* Sheath, with ligule, X2½. *c.* Spikelet, X1½. *d.* Glumes, X2½. *e.* Lemma, X2½.—forma *arkansanus.* *f.* Spikelet, X3½. *g.* Lemma, X2½.

5b. Elymus villosus Muhl. f. **arkansanus** (Scribn. & Ball) Fern. Rhodora 35:195. 1933. *Fig. 211f–g.*

Elymus arkansanus Scribn. & Ball, Bull. U.S.D.A. Div. Agrost. 24:45. 1901.

Elymus striatus var. *arkansanus* (Scribn. & Ball) Hitchcock, Rhodora 8:212. 1906.

Glumes scabrous; lemmas glabrous or scabrous.

RANGE: Maryland to Wyoming, south to Texas and North Carolina.

ILLINOIS DISTRIBUTION: Rare; known only from DuPage, Henry, and Stark counties.

6. Elymus canadensis L. Sp. Pl. 83. 1753. *Fig. 212.*

Elymus philadelphicus L. Cent. Pl. 1:6. 1755.

Elymus glaucifolius Muhl. in Willd. Enum. Pl. 1:131. 1809.

Elymus canadensis var. *glaucifolius* (Muhl.) Torr. Fl. N. & Midl. U. S. 1:137. 1824.

Elymus canadensis f. *crescendus* Ramaley, Minn. Bot. Stud. 1:114. 1894.

Elymus robustus Scribn. & Smith, Bull. U.S.D.A. Div. Agrost. 4:37. 1897.

Elymus brachystachys Scribn. & Ball, Bull. U.S.D.A. Div. Agrost. 24:47. 1901.

Elymus canadensis var. *robustus* (Scribn. & Smith) Mack. & Bush, Man. Fl. Jackson Co. Mo. 38. 1902.

Elymus crescendus (Ramaley) Wheeler, Minn. Bot. Stud. 3:106. 1903.

Elymus canadensis var. *brachystachys* (Scribn. & Ball) Farwell, Rep. Mich. Acad. Sci. 21:357. 1920.

Elymus glaucifolius crescendus (Ramaley) Bush, Am. Midl. Nat. 10:83. 1926.

Elymus canadensis f. *glaucifolius* Fern. Rhodora 35:191. 1933.
Densely cespitose perennial with culms to 2 m tall; sheaths glabrous or pubescent; blades flat, involute at the tip, thick, glabrous or scabrous or sparsely pilose, green or glaucous, 5–20 mm broad;

212. *Elymus canadensis* (Nodding Wild Rye). *a.* Inflorescence, X½. *b.* Sheath, with ligule, X2½. *c.* Spikelets, X2½. *d.* Glumes, X3½. *e.* Lemma, X3½. *f.* Lemma, X3½.

spikes dense, stiff, erect or somewhat nodding, 10–25 cm long, 1–2 cm broad; spikelets 2- to 7-flowered; glumes flat, 2- to 5-nerved, glabrous or scabrous or pubescent, 15–35 mm long, 0.5–2.0 mm broad; lemmas glabrous or scabrous or villous, 25–50 mm long (including the curving awns); paleas 8.5–11.0 mm long; 2n = 28 (Avdulov, 1928).

COMMON NAME: Nodding Wild Rye.

HABITAT: Woods; roadsides; dry prairies.

RANGE: Quebec to Alaska, south to California, Texas, and North Carolina.

ILLINOIS DISTRIBUTION: Common; in every county.

This common species is the most robust of the genus in Illinois.

Other distinguishing characters are its curving awns of the lemmas and its long paleas.

The lengthy synonymy is testimony to the variability of this species. Although various authors recognize several of these variations, every one is ill-defined and difficult to apply properly. Typical *E. canadensis* (including *E. philadelphicus*) is green with nodding, rather slender spikes and villous lemmas. Similar plants with glaucous blades have been known as *E. glaucifolius* (or var. *glaucifolius* or f. *glaucifolius*). Specimens with more or less glabrous lemmas have been referred to as *E. brachystachys* (or var. *brachystachys*). Plants with stout, nearly erect spikes, becoming more abundant westward, have been called *E. robustus* (or var. *robustus*) or *E. crescendus* (or f. *crescendus*). Although all the above variations have been observed in Illinois specimens, it does not appear possible to maintain these accurately as distinct taxa.

28. *Sitanion* RAF.

Cespitose perennials; blades flat; inflorescence spicate, dense; spikelets 2- to 4-flowered, usually paired at each joint of the rachis, disarticulating above the glumes; glumes 2, equal, setaceous, 1- to 3-nerved, awned; lemmas obscurely nerved, bifid at apex, sometimes awned; palea subequal to lemma.

The species which comprise this genus are highly variable, a fact which has resulted in disagreement among botanists as to the number of species.

Wilson (1963) is the last to study the genus comprehensively. Only the following species occurs in Illinois.

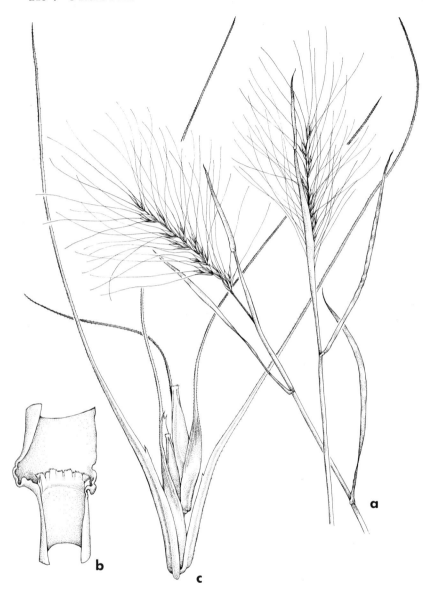

213. *Sitanion hystrix* (Squirrel-tail). *a*. Inflorescences, X½. *b*. Sheath, with ligule, X5. *c*. Spikelet, X2½.

1. **Sitanion hystrix** (Nutt.) J. G. Sm. Bull. U.S.D.A. Div. Agrost. 18:15. 1899. *Fig. 213.*

Aegilops hystrix Nutt. Gen. Pl. 1:86. 1818.

Tufted perennial to nearly 50 cm tall; sheaths glabrous or softly pubescent; blades narrow, sometimes involute, ascending to spreading, up to 3 (–5) mm broad; spike erect, mostly exserted, to about 8 cm long; spikelets two at each node, fertile; glumes very narrow to setiform, 1- (2-) nerved, tapering to a long, widely spreading awn; lemmas glabrous to slightly pubescent, with spreading awns up to 10 cm long.

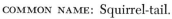

COMMON NAME: Squirrel-tail.

HABITAT: Along railroad (in Illinois).

RANGE: South Dakota to British Columbia, south to Texas and Missouri; Mexico; adventive in Illinois.

ILLINOIS DISTRIBUTION: Known from a single station in Mason County (cinders on abandoned railroad right-of-way, October 17, 1966, *R. T. Rexroat 10397*).

This species resembles *Hordeum jubatum* with which it was growing at the Illinois station. *Sitanion* has two perfect spikelets at each node, while in *H. jubatum*, there is only one perfect spikelet at the node.

The natural range of Squirrel-tail is west of Illinois.

29. *Hordeum* L. – Barley

Annuals or perennials; blades flat; inflorescence spicate, dense, bristly; spikelets 1- (2-) flowered, arranged 3 together at each joint of the rachis, the middle spikelet perfect, sessile, the outer usually imperfect, sessile or pedicellate, the entire group of 3 spikelets falling together; glumes setaceous (or widened at the base), usually awned, equal, rigid, situated in front of the spikelet; lemmas rounded on the back, that of the central spikelet indurate, obscurely nerved, long-awned, that of each of the lateral spikelets usually small and abortive.

A study of North American species of *Hordeum* is by Covas (1949).

KEY TO THE SPECIES OF Hordeum IN ILLINOIS

1. Lateral spikelets of each group of three pedicellate and sterile.
    2. Fertile spikelet subtended by awns less than 2 cm long; spikes less than 2 cm broad.

3. Two glumes of each group of 3 spikelets setiform, the other four glumes dilated at the base; awn of lemma of central spikelet 8–15 mm long_____1. *H. pusillum*

3. All glumes setiform; awn of lemma of central spikelet 5–6 mm long_____2. *H. brachyantherum*

2. Fertile spikelet subtended by awns exceeding 2 cm long; spikes 2–8 cm broad_____3. *H. jubatum*

1. Lateral spikelets of each group of three sessile and perfect.

4. Annual; spikes erect; glumes flat, about 1 mm broad; awns of lemmas 60–150 mm long, or absent and replaced by a three-lobed appendage_____4. *H. vulgare*

4. Perennial; spikes nodding; glumes setiform; awns of lemmas 15–35 mm long_____5. *H.* × *montanense*

1. **Hordeum pusillum** Nutt. Gen. Pl. 1:87. 1818. *Fig. 214.*

Hordeum riehlii Steud. Syn. Pl. Glum. 1:353. 1854.

Annual, sometimes decumbent at the base, with culms to 35 cm tall; sheaths scabrous; blades flat, scabrous, 2–6 mm broad; spikes erect, 2–7 cm long, 1.0–1.5 cm broad, green; lateral spikelets pedicellate, abortive, the first glume dilated above the base and with an awn 8–15 mm long, the second glume setiform, the lemma reduced and awn-tipped; central spikelet sessile, fertile, both glumes dilated above the base, 12–15 mm long, awned, the lemma with an awn 8–15 mm long.

COMMON NAME: Little Barley.

HABITAT: Fields, waste ground.

RANGE: Throughout the United States.

ILLINOIS DISTRIBUTION: Common in the southern three-fourths of the state; rare in the northern one-fourth.

This is the smallest species of *Hordeum* in Illinois, as well as one of the most abundant. It flowers from early May to early July. The pedicellate lateral spikelets relate it to *H. jubatum* and *H. brachyantherum*, but both these species have glumes which are all setiform. *Hordeum pusillum* has more narrow spikes and is of shorter stature than any other *Hordeum* in Illinois, although it is approached in both characters by *H. brachyantherum*.

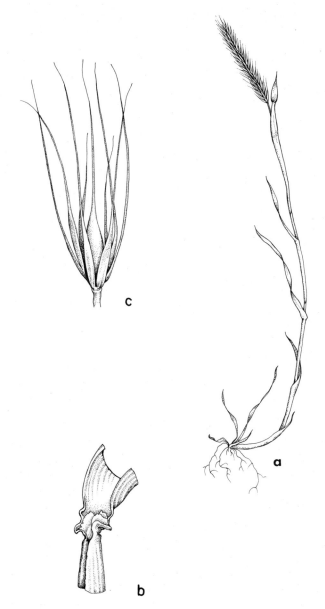

214. Hordeum pusillum (Little Barley). *a.* Habit, X½. *b.* Sheath, with ligule, X3½. *c.* Spikelets, X5.

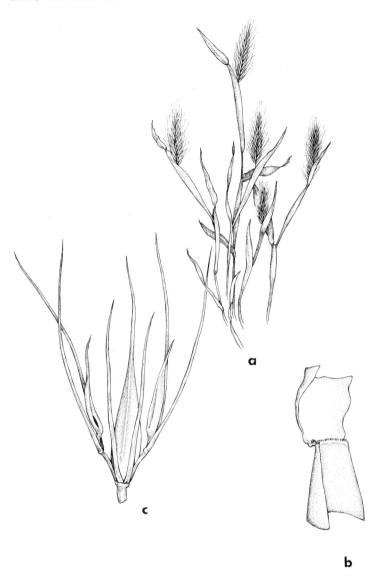

215. *Hordeum brachyantherum* (Meadow Barley). *a.* Upper parts of plants, X½. *b.* Sheath, with ligule, X5. *c.* Spikelets, X5.

2. Hordeum brachyantherum Nevski, Acta Inst. Bot Acad. Sci. U.R.S.S.I. 2:61. 1936. *Fig. 215.*

Hordeum boreale Scribn. & Smith, Bull. U.S.D.A. Div. Agrost. 4:24. 1897, non Gandog. (1881).

Hordeum nodosum L. var. *boreale* (Scribn. & Smith) Hitchc. Am. Jour. Bot. 21:134. 1934.

Tufted perennial with more or less upright culms to 50 cm tall (in Illinois); sheaths glabrous or softly hairy; blades flat, 3–8 mm broad; spikes mostly erect or suberect, 2–7 cm long, 0.5–1.2 cm broad, often purplish; lateral spikelets pedicellate, abortive; central spikelet sessile, fertile; all glumes setiform; lemma of central spikelet 5–6 mm long, awned, the awn 5–6 mm long.

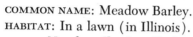

COMMON NAME: Meadow Barley.

HABITAT: In a lawn (in Illinois).

RANGE: Newfoundland to Alaska, south to California and New Mexico; adventive in a few eastern states.

ILLINOIS DISTRIBUTION: Known from a single collection (Jackson Co.: 3½ miles west of Carbondale on the John Voigt property, May 30, 1965, *J. W. Voigt s. n.*).

The Meadow Barley, a native grass across northern North America, is sparingly adventive in the eastern United States where its seeds are probably mixed with those of other lawn grasses. The small stature and slender erect spikes resemble those of *H. pusillum* but, in *H. pusillum,* four of the six glumes in each group of three spikelets are dilated at the base.

3. Hordeum jubatum L. Sp. Pl. 85. 1753. *Fig. 216.*

Densely tufted perennial, sometimes decumbent at the base, with culms to 60 cm tall; sheaths scabrous; blades scabrous, 2–5 mm broad; spikes nodding, 5–10 cm long, often nearly as broad, pale green or purple; lateral spikelets pedicellate, reduced to 1–3 spreading, scabrous awns; central spikelet sessile, fertile, the glumes setiform, equal, spreading, 25–65 mm long, the lemma 5–8 mm long with a scabrous awn 25–60 mm long; 2n = 14 (Tanzi, 1925), 28 (Aase & Powers, 1926).

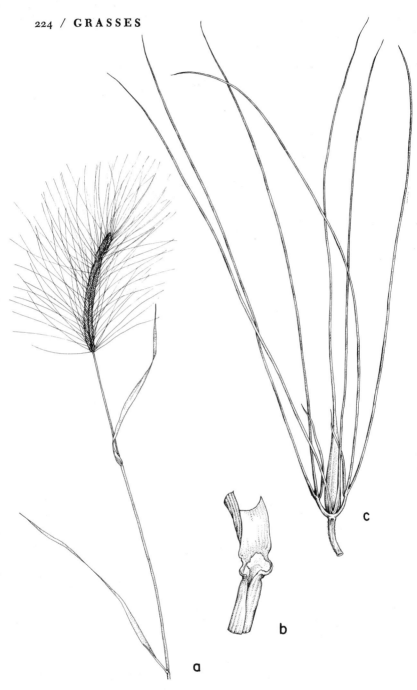

216. *Hordeum jubatum* (Squirrel-tail Grass). *a.* Inflorescence, X½. *b.* Sheath, with ligule, X2½. *c.* Spikelets, X3½.

COMMON NAME: Squirrel-tail Grass.

HABITAT: Fields.

RANGE: Throughout the United States; adventive in Europe and Asia.

ILLINOIS DISTRIBUTION: Common in the northern one-half of the state; occasional in the southern one-half.

The long-awned spikelets make this one of the most attractive grasses in Illinois. The flowers are produced from June to early September.

4. Hordeum vulgare L. Sp. Pl. 84. 1753. *Fig. 217.*

Annual to 1.2 m tall; blades flat (3–) 5–15 mm broad, auriculate at the base; spikes erect, dense, to 10 cm long (excluding the awns); lateral and central spikelets sessile, perfect; glumes flat, about 1 mm broad, divergent, nerveless, pubescent, awned; lemmas about 10 mm long, with a straight awn 60–150 mm long or with the awn replaced by a three-lobed structure; $2n = 14$ (Kihara, 1924).

A commonly cultivated species, with two varieties sometimes adventive but rarely persisting in Illinois.

1. Lemmas awned_____4a. *H. vulgare* var. *vulgare*
1. Lemmas three-lobed at the apex, awnless_____
_____4b. *H. vulgare* var. *trifurcatum*

4a. Hordeum vulgare L. var. **vulgare.**

Lemmas awned.

COMMON NAME: Common Barley.

This variety is found occasionally along country roads in Illinois. It is probably native to Asia.

4b. Hordeum vulgare L. var. **trifurcatum** (Schlecht.) Alefeld, Landw. Fl. 341. 1866.

Hordeum coeleste var. *trifurcatum* Schlecht. Linnaea 11:543. 1837.

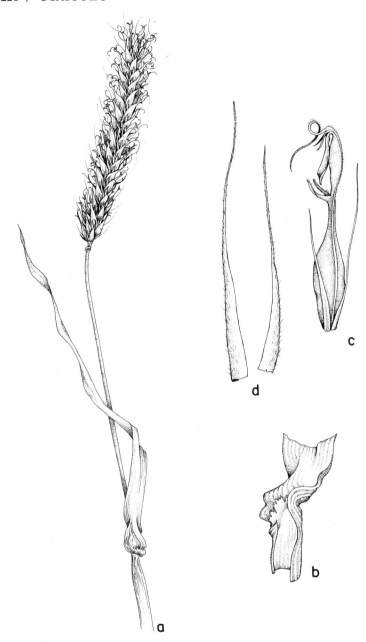

217. *Hordeum vulgare* (Common Barley). *a.* Inflorescence, X½. *b.* Sheath, with ligule, X2½. *c.* Spikelet, X3½. *d.* Glumes, X6.

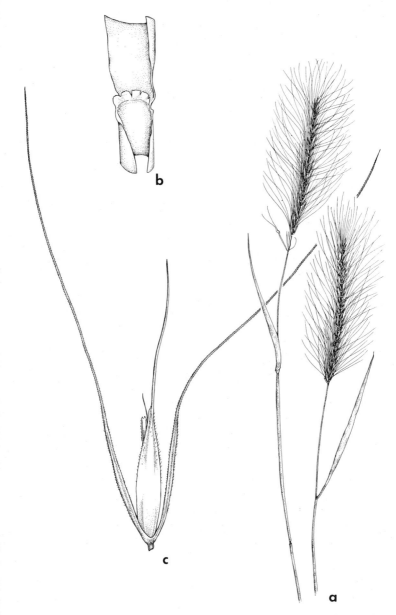

218. *Hordeum × montanense* (Barley). *a.* Inflorescences, X½. *b.* Sheath, with ligule, X5. *c.* Spikelet, X5.

Hordeum trifurcatum (Schlecht.) Wender, Flora 26:233. 1843. Lemmas three-lobed at the apex, awnless.

> COMMON NAME: Pearl Barley.
> This variety is less common than the preceding in Illinois (not mapped).

5. **Hordeum × montanense** Scribn. in Beal, Grasses N. Am. 2:644. 1896. *Fig. 218.*

Hordeum pammeli Scribn. & Ball, Suppl. Rept. Iowa Geol. Surv. 1903:335. 1905.

Perennial, sometimes decumbent at the base, to 1 m tall; sheaths glabrous; blades flat, scabrous, 5–8 mm broad; spikes nodding, 6–17 cm long, 2–3 cm broad; lateral and central spikelets sessile, fertile, the lateral 1-flowered, the central 2-flowered and with a rudimentary third floret; glumes setiform, 20–35 mm long (including the awn); lemmas 7–8 mm long, the awn 15–35 mm long.

> COMMON NAME: Barley.
> HABITAT: Prairies and roadsides (in Illinois).
> RANGE: Illinois, west through Iowa, South Dakota, Wyoming, and Montana.
> ILLINOIS DISTRIBUTION: Not common; known only from Marshall, Peoria, and Stark counties.
> This species is a reputed hybrid between *H. jubatum* and *Elymus virginicus*. It flowers during June and July in Illinois.

Hordeum × montanense differs from *H. jubatum* by having the lateral spikelets sessile and by having usually broader blades.

30. × Agrohordeum G. CAMUS EX ROUSSEAU

This hybrid genus formed between *Agropyron* and *Hordeum* shares characters of both parent genera while looking remarkably like the genus *Elymus*.

Rousseau (1952) validated the hybrid generic name × *Agrohordeum*.

Only the following species of this genus has been found in Illinois.

1. **Agrohordeum × macounii** (Vasey) Lepage, Nat. Can. 79:242. 1952. *Fig. 219.*

219. Agrohordeum × macounii (Macoun's Wild Rye). *a.* Upper part of plant, X½. *b.* Sheath, with ligule, X5. *c.* Spikelet, X7½. *d.* Lemma, X7½.

Elymus macounii Vasey, Bull. Torrey Club 13:119. 1886.
Tufted perennials without rhizomes, with culms nearly 1 m tall; sheaths mostly glabrous; blades firm, flat or becoming involute, scabrous on both surfaces, up to 5 mm wide; spikes slender, erect or slightly nodding, to 12 cm long, to 0.5 cm broad; spikelets (1-) 2-flowered, appressed, the uppermost spikelets paired, the lowermost apparently solitary; glumes linear-setaceous, scabrous, 3-nerved; lemmas to 1 cm long, scabrous near the apex, with an awn up to 2 cm long.

COMMON NAME: Macoun's Wild Rye.
HABITAT: Along railroad (in Illinois).
RANGE: Manitoba to Alaska, south to California, New Mexico, and Iowa; adventive in Illinois.
ILLINOIS DISTRIBUTION: Cook Co.: Cicero, along Santa Fe Railroad W of Cicero Avenue, July 1, 1963, *F. Swink.* Experimental evidence by Boyle and Holmgren (1955) and by Gross (1960) shows that this species is a hybrid formed between *Agropyron trachycaulum* (Link) Malte and *Hordeum jubatum* L., an idea first put forth years earlier by Stebbins.

Vasey originally described this species as an *Elymus*, and most authors to date have continued to treat it as such. The lowermost spikelet being borne singly distinguishes *A.* × *macounii* from *Elymus*.

31. Agropyron GAERTN. – Wheat Grass

Perennials; blades flat or involute; inflorescence spicate, erect; spikelets several-flowered, solitary and placed flatwise at each joint of the rachis, disarticulating above the glumes; glumes 2, subequal, conspicuously 1- to 7-nerved, acuminate or short-awned, shorter than the spikelet; lemmas rounded on the back (3-) 5- to 7-nerved, acute or awned.

This genus is more abundant in the western United States where several species serve as excellent forage grasses. There is strong evidence which indicates an extremely close relationship between *Agropyron* and *Elymus*.

KEY TO THE SPECIES OF Agropyron IN ILLINOIS

1. Lemmas 5–7 mm long; glumes 2–5 mm long; spikelets pectinately arranged; tufted plants.
    2. Blades flat; spikelets 8–12 mm long; glumes abruptly tapering to

the 2–3 mm long awn_____1. *A. desertorum*
2. Blades involute (at least when dry); spikelets 5–7 mm long;
  glumes gradually tapering to the 2–5 mm long awn_____
  _____2. *A. cristatum*
1. Lemmas 8–25 mm long; glumes 8–18 mm long; spikelets not pectinately arranged; tufted or rhizomatous plants.
  3. Tufted plants; spikelets at maturity (and on herbarium speci-
    mens) readily breaking up into individual florets when touched.
    4. Awn of lemmas 10–30 mm long; spikes dense; glumes 12–18
      mm long_____3. *A. subsecundum*
    4. Awn of lemmas absent or up to 2 mm long; spikes more
      slender; glumes 8–12 mm long_____4. *A. trachycaulum*
  3. Rhizomatous plants; spikelets at maturity falling in their entirety.
    5. Blades flat, 5–10 mm broad; sheaths pubescent (in Illinois)__
      _____5. *A. repens*
    5. Blades involute when dry, 2–5 mm broad; sheaths glabrous
      (in Illinois)_____6. *A. smithii*

1. **Agropyron desertorum** (Fisch.) Schult. Mantissa 2:412.
1824. *Fig. 220.*

Triticum desertorum Fisch. ex Link, Enum. Pl. 1:97. 1821.
Densely tufted perennial to 1 m tall; lower sheaths spreading
hirsute, the upper more or less glabrous; blades 2–5 mm broad,
flat; spikes 5–9 cm long; spikelets 8–12 mm long, (3-) 5- to 7-flow-
ered; glumes firm, glabrous to sparsely ciliate along the keel, subequal, 2–5 mm long, abruptly tapering to a 2–3 mm long awn;
lemmas firm, glabrous to sparsely ciliate along the keel, 5–7 mm
long, the awn 2–3 mm long, bent to one side.

COMMON NAME: Wheat Grass.
HABITAT: Railroad yards (in Illinois).
RANGE: Native of Russia; introduced in the western
United States; adventive in New York and Illinois.
ILLINOIS DISTRIBUTION: Rare; known only from Cook and
JoDaviess counties; first collected in Illinois in 1956
(Milwaukee Road classification yard west of Franklin
Park, August 9, 1956, *J. W. Thieret 2295*).
This and the following species are readily distinguished
  from the other *Agropyrons* in Illinois because of their
pectinately arranged spikelets. *Agropyron desertorum* is similar
to *A. cristatum*, but has flat blades and slightly longer spikelets.

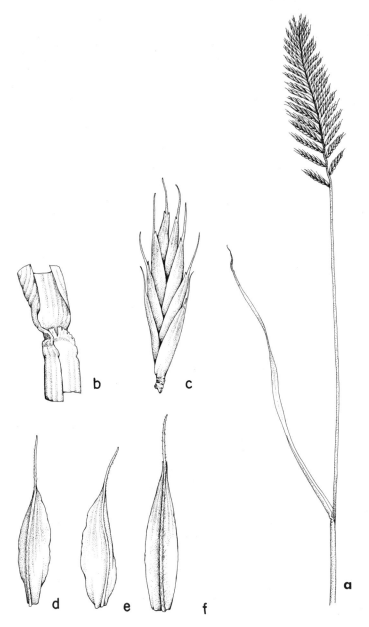

220. *Agropyron desertorum* (Wheat Grass). *a.* Inflorescence, X½. *b.* Sheath, with ligule, X2½. *c.* Spikelet, X5. *d.* First glume, X6. *e.* Second glume, X6. *f.* Lemma, X6.

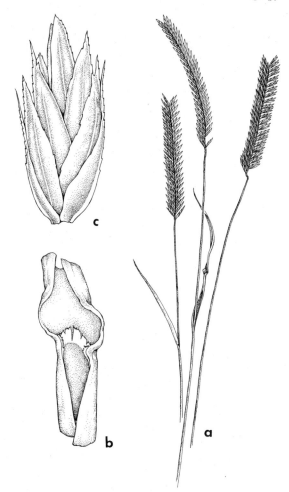

221. Agropyron cristatum (Crested Wheat Grass). *a.* Upper part of plants, X½. *b.* Sheath, with ligule, X2½. *c.* Spikelet, X7½.

2. Agropyron cristatum (L.) Gaertn. Nov. Comm. Petrop. 14: 540. 1770. *Fig. 221.*

Bromus cristatus L. Sp. Pl. 78. 1753.

Densely tufted perennial to nearly 1 m tall; sheaths pubescent to glabrous; blades involute at maturity; spikes 2–7 cm long; spikelets 5–7 mm long, 3- to 8-flowered; glumes firm, glabrous to ciliate along the keel, 2–5 mm long, gradually tapering into the 2–5 mm long curved awns; lemmas firm, glabrous to ciliate along the keel,

222. *Agropyron subsecundum* (Bearded Wheat Grass). *a.* Inflorescences, X½. *b.* Sheath, with ligule, X5. *c.* Spikelet, X10.

5–7 mm long, the awn 2–5 mm long, curved; 2n = 14, 28 (Hartung, 1946).

COMMON NAME: Crested Wheat Grass.

HABITAT: On a mine spoilbank and along a railroad (in Illinois).

RANGE: Native of Russia; rarely adventive in New York, North Dakota, and Illinois.

ILLINOIS DISTRIBUTION: Rare; known only from Fulton County (growing on a mine spoilbank with *Bouteloua curtipendula*, 1½ miles W of Rapatee, July 26, 1950, *J. R. Fuelleman*) and JoDaviess County (5 miles southwest of Galena along railroad track near Mississippi River, July 15, 1966, *R. P. Wunderlin & W. Chapman s.n.*). The State Department of Highways has also planted this species along the Kennedy Expressway in Cook County.

3. **Agropyron subsecundum** (Link) Hitchcock, Amer. Journ. Bot. 21:131. 1934. *Fig. 222.*

Triticum subsecundum Link, Hort. Berol. 2:190. 1833.

Agropyron richardsoni Schrad. Linnaea 12:467. 1838, *in synon.*

Agropyron unilaterale Cassidy, Bull. Colo. Agr. Exp. Sta. 12:63. 1890.

Agropyron caninum var. *unilaterale* (Cassidy) Vasey, Contrib. U. S. Nat. Herb. 1:279. 1893.

Agropyron caninum f. *glaucum* Pease & Moore, Rhodora 12:71. 1910.

Agropyron trachycaulum var. *unilaterale* (Cassidy) Malte, Ann. Rep. Can. Nat. Mus. 1930:46. 1932.

Agropyron trachycaulum var. *glaucum* (Pease & Moore) Malte, Ann. Rep. Can. Nat. Mus. 1930:47. 1932.

Tufted perennial with culms to 1 m tall; sheaths usually glabrous; blades flat, 3–8 mm broad, green or glaucous; spikes 5–20 cm long; spikelets 12–18 mm long, acuminate or awn-tipped; lemmas firm, obscurely 5-nerved, 10–25 mm long, the awn 10–30 mm long; 2n = 28 (Hartung, 1946).

COMMON NAME: Bearded Wheat Grass.

HABITAT: Woodlands, fields.

RANGE: Ontario to British Columbia, south to Oregon, Colorado, and Illinois.

ILLINOIS DISTRIBUTION: Rare; limited to the extreme northern counties.

This species may have either green or glaucous leaves. Although several recent authors consider it to be a variety of *A. trachycaulum,* the differences in the length of the glumes and awns of the lemmas seem to justify specific segregation.

This species flowers in July and August.

4. **Agropyron trachycaulum** (Link) Malte, Ann. Rep. Can. Nat. Mus. 1930:42. 1932. *Fig. 223.*

Triticum pauciflorum Schwein. in Keat. Narr. Exped. St. Peter's River 2:383. 1824, *non A. pauciflorum* Schur (1859).

Triticum trachycaulum Link, Hort. Berol. 2:189. 1833.

Agropyron tenerum Vasey, Bot. Gaz. 10:258. 1885.

Agropyron pauciflorum (Schw.) Hitchcock, Am. Journ. Bot. 21:132. 1934.

Tufted perennial with culms to 1 m tall; sheaths usually glabrous; blades flat, 2–4 (–8) mm broad, green or glaucous; spikes 5–25 cm long, more slender than in *A. subsecundum;* spikelets 12–15 mm long (excluding the awns), 2- to 7-flowered; glumes subequal, keeled, 8–12 mm long, acuminate or awn-tipped; lemmas firm, obscurely 5-nerved, 10–25 mm long, the awn absent or only up to 2 mm long.

COMMON NAME: Slender Wheat Grass.

HABITAT: Fields and waste ground; railroad tracks.

RANGE: Labrador to Alaska, south to California, Illinois, and West Virginia; probably adventive in Illinois.

ILLINOIS DISTRIBUTION: Rare; known from four extreme northern counties.

This species differs from *A. subsecundum* in its very short or absent awns of the lemma and in its slightly shorter glumes. The tip of each lemma does not reach the base of the next spikelet above.

The earliest reports of this species from Illinois were made under the binomial *A. tenerum.*

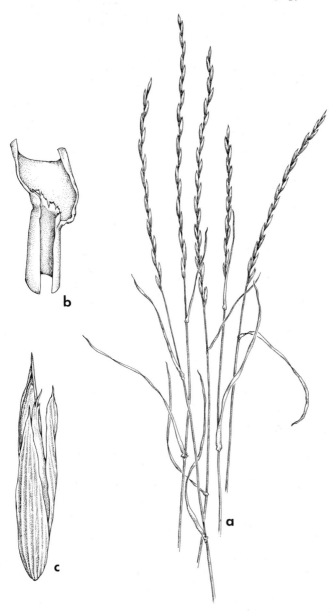

223. Agropyron trachycaulum (Slender Wheat Grass). *a.* Upper parts of plant, X½. *b.* Sheath, with ligule, X2½. *c.* Spikelet, X6.

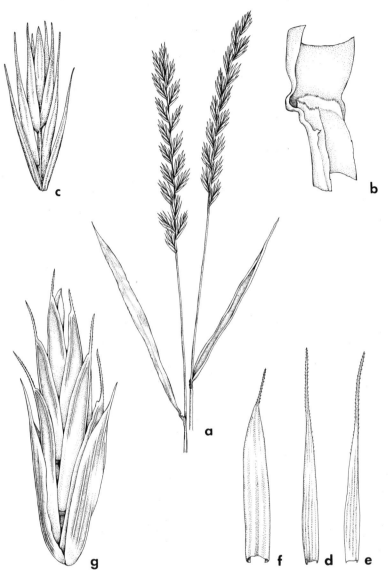

224. *Agropyron repens* (Quack Grass).—var. *repens*. *a*. Inflorescence, X½. *b*. Sheath, with ligule, X4. *c*. Spikelet, X3. *d*. First glume, X5. *e*. Second glume, X5. *f*. Lemma, X5.—var. *aristatum*. *g*. Spikelet, X5.

5. **Agropyron repens** (L.) Beauv. Ess. Agrost. 102. 1812.

Triticum repens L. Sp. Pl. 86. 1753.

Rhizomatous perennial with culms to 1 m tall; sheaths pubescent; blades flat, sparsely pubescent above, green or glaucous, 5–10 mm broad; spikes 6–17 cm long; spikelets 10–18 mm long, 4- to 8-flowered; glumes narrowly oblong to lanceolate, acuminate or awn-tipped, 8–14 mm long; lemmas narrowly oblong to lanceolate, glabrous or scabrous, 8–10 (–12) mm long, awnless or with an awn up to 10 mm long; 2n = 28, 42 (Avdulov, 1931).

Agropyron repens is similar to *A. smithii*, but differs in its pubescent sheaths and its flat, broader blades.

Two forms occur in Illinois.

1. Lemmas awnless_____5a. *A. repens* f. *repens*
1. Lemmas with an awn up to 10 mm long__5b. *A. repens* f. *aristatum*

5a. **Agropyron repens** (L.) Beauv. f. **repens** *Fig. 224a–f.*
Lemmas awnless.

COMMON NAME: Quack Grass.

HABITAT: Fields and waste ground.

RANGE: Native of Europe and Asia; adventive throughout most of North America.

ILLINOIS DISTRIBUTION: Common in the northern three-fourths of the state; rare in the southern one-fourth.

This form, the more abundant in Illinois, flowers in June and July.

This taxon is a common weed in the northern half of Illinois. Its creeping rhizomes make it difficult to exterminate once it has become established. Quack Grass may be utilized in hay production.

5b. **Agropyron repens** (L.) Beauv. f. **aristatum** (Schum.) Holmb. Skand. Fl. 2:274. 1926. *Fig. 224g.*

Triticum repens var. *aristatum* Schum. Enum. Pl. Saell. 2:38. 1803.

Triticum vaillantianum Wulf. & Schreb. in Schweig. & Korte, Spec. Fl. Erlang. 1:143. 1804.

Agropyron repens var. *subulatum* f. *vaillantianum* (Wulf. & Schreb.) Fern. Rhodora 35:184. 1933.

Lemmas with an awn up to 10 mm long.

 ILLINOIS DISTRIBUTION: Rare; Cook County.

6. Agropyron smithii Rydb. Mem. N. Y. Bot. Gard. 1:64. 1900. (Feb.)

Agropyron occidentale Scribn. Circ. U.S.D.A. Div. Agrost. 27:9. 1900. (Dec.)

Rhizomatous perennial with culms to 90 cm tall; sheaths glabrous; blades involute when dry, scabrous or villous above, glaucous, 2–5 mm broad; spikes 7–15 cm long; spikelets 12–22 mm long, 6- to 12-flowered; glumes narrowly lanceolate, acute to long-acuminate, 9–14 mm long; lemmas narrowly lanceolate, glabrous or scabrous or pubescent only at the base, rarely short-pilose, 10–14 mm long, awnless or with an awn to 1.5 mm long; 2n = 28, 56 (Hartung, 1946).

Two varieties occur in Illinois.

1. Lemmas glabrous, scabrous, or pubescent near base_____ _____6a. A. *smithii* var. *smithii*
1. Lemmas short-pilose throughout_____6b. A. *smithii* var. *molle*

6a. Agropyron smithii Rydb. var. **smithii** *Fig. 225a–f.*

Lemmas glabrous, scabrous, or pubescent near base.

 COMMON NAME: Western Wheat Grass.
HABITAT: Along railroads (in Illinois).
RANGE: Western United States; adventive eastward.
ILLINOIS DISTRIBUTION: Occasional in the northern two-thirds of the state; rare in the southern one-third.
The leaves and culms have a silvery or bluish cast, while the leaves are unique in their involute margins.
This is an important pasture grass of the Great Plains. Occasional specimens are found with the spikes bearing two spikelets at a node.

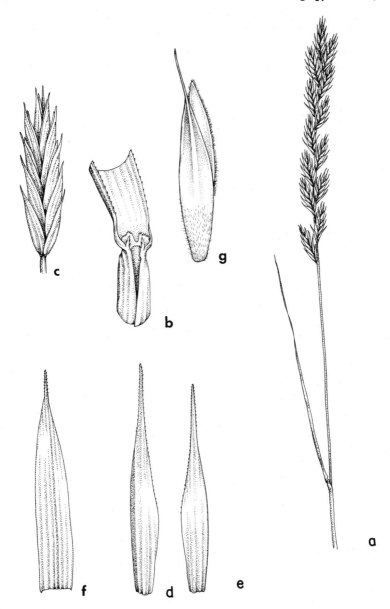

225. Agropyron smithii (Western Wheat Grass).—var. *smithii.* *a.* Inflorescence, X½. *b.* Sheath, with ligule, X4. *c.* Spikelet, X2½. *d.* First
glume, X7½. *e.* Second glume, X7½. *f.* Lemma, X7½.—var. *molle.* *g.*
Lemma, X7½.

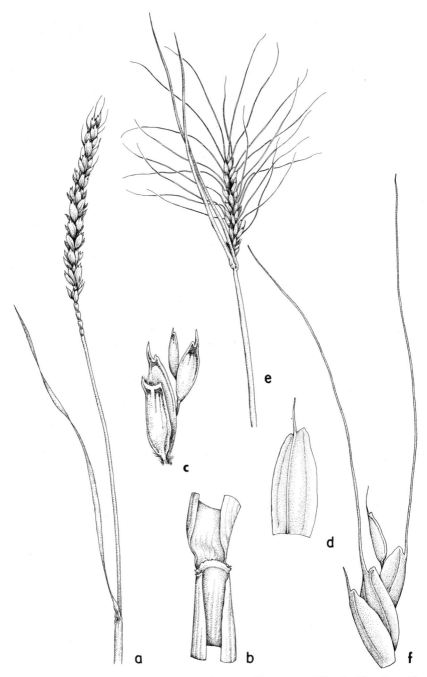

226. *Triticum aestivum* (Wheat). *a.* Inflorescence, X½. *b.* Sheath, with ligule, X4. *c.* Spikelets, X3½. *d.* Glume, X5. *e.* Inflorescence (bearded form), X½. *f.* Spikelets (bearded form), X½.

6b. Agropyron smithii Rydb. var. **molle** (Scribn. & Smith)
Jones, Contr. West. Bot. 14:18. 1912. *Fig. 225g.*
Agropyron spicatum var. *molle* Scribn. & Smith, Bull. U.S.D.A.
Div. Agrost. 4:33. 1897.
Agropyron molle (Scribn. & Smith) Rydb. Mem. N. Y. Bot.
Gard. 1:65. 1900.
Lemmas short-pilose throughout.

HABITAT: Along railroads (in Illinois).
RANGE: Western United States; adventive in Illinois.
ILLINOIS DISTRIBUTION: Cook, DuPage, and Will counties.
Specimens previously called *Agropyron dasystachyum*
from Illinois actually are *A. smithii* var. *molle.*
The specimens from Will County (*S.F. Glassman 4312a,
4313*) have 2–4 spikelets at each node.

32. *Triticum* L. – Wheat

Annuals; blades flat; inflorescence spicate, dense; spikelets 2- to
5-flowered, solitary, placed flatwise at each joint of the rachis, disarticulating above the glumes or the entire spikelet disjointing
from the plant at maturity; glumes rigid, keeled, conspicuously
nerved, asymmetrical, toothed, mucronate, or awned at the apex;
lemmas keeled, asymmetrical, mucronate or awned.

The recent tendency among students of grasses is to combine
the genus *Aegilops* with *Triticum,* a view followed in this work.
The major difference between the two is in the manner of disarticulation of the spikelets.

KEY TO THE SPECIES OF Triticum IN ILLINOIS

1. Spikelets compressed, disarticulating above the glumes; joints of
   rachis not swollen; blades 10–20 mm broad_____1. *T. aestivum*
1. Spikelets cylindrical, falling in their entirety; joints of rachis swollen; blades 2–3 mm broad_____2. *T. cylindricum*

1. **Triticum aestivum** L. Sp. Pl. 85. 1753. *Fig. 226.*

Triticum vulgare Villars, Hist. Pl. Dauph. 2:153. 1787.
Culms to 1.2 m tall; blades 10–20 mm broad; spike 5–12 cm long;
spikelets highly variable in size and pubescence; lemmas awnless
or with awns up to 8 cm long.

COMMON NAME: Wheat; Bearded Wheat (with the long awns).

HABITAT: Waste ground.

Wheat occurs sporadically in fields and along roadsides. It is doubtful that it is ever truly persistent.

Many cultivated varieties may be found in Illinois. For the names of some of these, see Hitchcock (1950), page 245.

2. Triticum cylindricum (Host) Ces. Pass. & Gib. Comp. Fl. Ital. 86. 1867. *Fig. 227.*

Aegilops cylindrica Host, Icon. Gram. Austr. 2:6. 1802.

Tufted annual, branching from the base, with the culms to 60 cm tall; blades 2–3 mm broad; spikes cylindric, 3–10 cm long; spikelets 8–10 mm long, 2- to 5-flowered, glabrous or hispid; glumes excentrically keeled, the keel prolonged into an awn less than 1 cm long, with one of the lateral nerves extending into a short tooth; lemmas mucronate, the uppermost with scabrous awns to 50 mm long.

COMMON NAME: Jointed Goat Grass.

HABITAT: Waste ground.

RANGE: Native of Europe; introduced primarily in the western United States.

ILLINOIS DISTRIBUTION: Occasional; scattered throughout the state.

Goat Grass flowers during the summer. It is one of the strangest appearing grasses in the state. In certain regions of the United States, this species may become a weed in wheat fields.

33. *Secale* L. – Rye

Annuals; blades flat; inflorescence spicate, dense; spikelets 2-flowered, solitary, placed flatwise at each joint of the rachis, flattened, disarticulating above the glumes; glumes 2, rigid, shorter than the spikelets; lemmas excentrically keeled, 5-nerved, awned.

1. Secale cereale L. Sp. Pl. 84. 1753. *Fig. 228.*

Culms branching from the base, to 1.5 m tall; blades 8–20 mm broad; spikes 8–15 cm long, somewhat nodding; glumes linear-subulate, 1-nerved; lemmas lance-subulate, the awn up to 8 cm long.

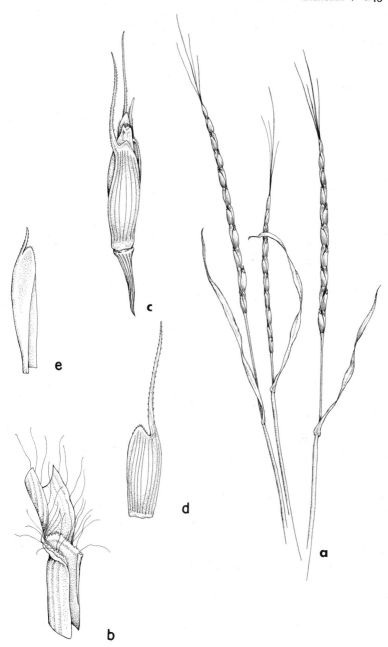

227. *Triticum cylindricum* (Jointed Goat Grass). *a.* Inflorescences, X½. *b.* Sheath, with ligule, X4. *c.* Spikelet, X3. *d.* Glume, X3. *e.* Lemma, X3.

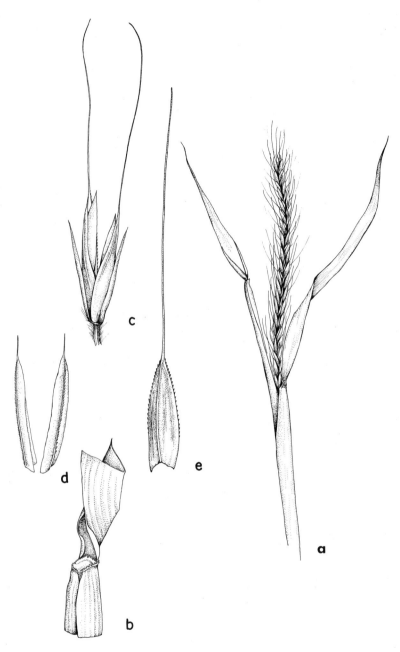

228. *Secale cereale* (Rye). *a.* Inflorescence, X½. *b.* Sheath, with ligule, X2½. *c.* Spikelet, X3. *d.* Glumes, X4. *e.* Lemma, X4.

COMMON NAME: Rye.

HABITAT: Waste ground.

As with Wheat, this species probably never becomes persistent in Illinois. It is planted frequently by the highway department along new road rights-of-way. It flowers from mid-May until late July.

This species is highly susceptible to infection by the fungus, *Claviceps purpurea,* or ergot, which causes the production of purplish-black sclerotia throughout the inflorescence.

Tribe *Meliceae*

Mostly tall perennials (in Illinois); inflorescence a panicle or raceme; spikelets several-flowered, disarticulating above the glumes; glumes subequal or distinctly unequal, awnless; lemmas 5- to 13-nerved, awned to awnless.

The three genera which comprise this tribe in Illinois (*Melica, Glyceria, Schizachne*) have traditionally been placed in the Festuceae. They differ from other festucaceous genera such as *Bromus, Festuca,* and *Poa* by the virtual absence of silica cells in the leaf epidermis.

34. *Melica* L. – Melic Grass

Perennials; sheaths closed; blades flat, soft; inflorescence paniculate; spikelets 2- to several-flowered, disarticulating above the glumes; glumes 2, unequal, chartaceous, prominently nerved, a little shorter than the spikelets; lemmas rounded on the back, prominently nerved, awnless (in the Illinois species), the upper 2 or 3 smaller and sterile.

This genus, composed of handsome woodland or prairie species (in Illinois), is distinguished from other genera with several-flowered spikelets (except *Schizachne*) by the presence of sterile or staminate lemmas above the fertile ones; from *Schizachne* it differs in having a glabrous callus on the lemmas.

KEY TO THE SPECIES OF Melica IN ILLINOIS

1. Cauline leaves 3–4, 2–5 mm broad; sheaths scabrous; glumes nearly equal in length; first glume oblong, at least twice as long as broad; fertile lemmas usually 2_____1. *M. mutica*
1. Cauline leaves 5–8, 5–12 mm broad; sheaths glabrous; glumes unequal in length; first glume ovate, less than twice as long as broad; fertile lemmas usually 3_____2. *M. nitens*

1. Melica mutica Walt. Fl. Carol. 78. 1788. *Fig. 229.*

Loosely tufted perennial from knotty rhizomes; culms wiry, to 1 m tall; sheaths scabrous; blades 3–4 per culm, flat, 2–5 mm broad; inflorescence 10–20 cm long, ascending; spikelets 7–10 mm long, with 2 fertile florets, pedicellate; glumes nearly equal in length, oblong, 6.5–9.0 mm long, at least twice as long as broad; fertile lemmas obtuse, 7- to 13-nerved, 7–10 mm long, scaberulous, the sterile lemmas about 2 mm long; 2n = 18 (unpublished data).

COMMON NAME: Two-flowered Melic Grass.

HABITAT: Rocky woodlands.

RANGE: Maryland to Iowa, south to Texas and Florida.

ILLINOIS DISTRIBUTION: Occasional; scattered throughout the state. This species is still a conspicuous member of the grass flora of southern Illinois, although it is becoming rare or even extinct in the northernmost counties. Glassman (1964) reports no collection of it from the Chicago region since 1899.

This graceful woodland grass flowers from early May to late June. It is one of the most attractive grasses in the state. The fewer, narrower leaves and broader, scabrous glumes distinguish this species from *M. nitens*.

2. Melica nitens (Scribn.) Nutt. ex Piper, Bull. Torrey Club 32:387. 1905. *Fig. 230.*

Melica scabra Nutt. Trans. Amer. Phil. Soc. 5:148. 1837, non HBK. (1816).

Melica diffusa var. *nitens* Scribn. Proc. Acad. Nat. Sci. Phil. 1885:44. 1885.

Loosely tufted perennial from short rhizomes; culms wiry, 1 to 1.5 m tall; sheaths glabrous; blades 5–8 per culm, flat, 5–15 mm broad; inflorescence 10–25 cm long, ascending; spikelets 9–12 mm long, with 3 fertile florets, pedicellate; first glume ovate, 5–8 mm long; second glume ovate, 6–9 mm long; fertile lemmas acute, 7- to 13-nerved, 7–9 mm long, scabrous, the sterile lemmas about 2 mm long; 2n = 18 (unpublished data).

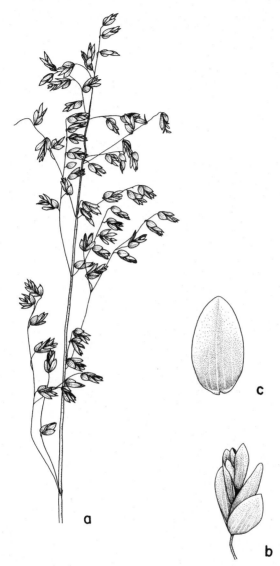

229. *Melica mutica* (Two-flowered Melic Grass). *a.* Inflorescence, X½. *b.* Spikelet, X2. *c.* First glume, X4.

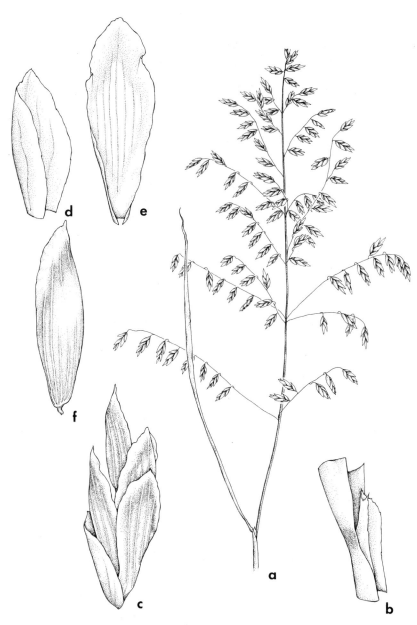

230. *Melica nitens* (Three-flowered Melic Grass). *a.* Inflorescence, X½.
b. Sheath, with ligule, X2½. *c.* Spikelet, X6. *d.* First glume, X7½. *e.*
Second glume, X7½. *f.* Lemma, X7½.

COMMON NAME: Three-flowered Melic Grass.

HABITAT: Rocky woods, prairies.

RANGE: Pennsylvania to Minnesota, south to Texas and Virginia.

ILLINOIS DISTRIBUTION: Occasional; throughout the state. Apparently a little more common than *M. mutica* but, as in the former, not collected from the Chicago region in over half a century, with the exception of a collection made from one large colony in Grundy County near Mazonia.

The flowers appear from mid-May to early July.

In comparison with *Melica mutica, M. nitens* generally grows taller and has glabrous sheaths, more blades per culm, wider blades, somewhat longer spikelets, and acute lemmas.

35. *Glyceria* R. BR. – Manna Grass

Perennials; sheaths usually closed; blades flat; inflorescence paniculate; spikelets several-flowered, disarticulating above the glumes; glumes 2, unequal, shorter than the spikelets; lemmas rounded on the back, distinctly 5- to 9-nerved, awnless; lodicules united; style present.

Chief apparent differences separating this genus from *Puccinellia* are the united lodicules of *Glyceria,* the distinctly nerved lemmas, and the usually closed sheaths. The plant usually known as *G. pallida* now is considered to be a species of *Puccinellia*.

A cytotaxonomic study of the genus has been made by Church (1949).

KEY TO THE SPECIES OF Glyceria IN ILLINOIS

1. Spikelets at least 10 mm long; sheaths compressed.
    2. Principal leaves 2–5 mm broad; lemmas shining, scabrous only on the nerves; pedicels very slender, all one-fourth to two-thirds the length of the spikelets_____1. *G. borealis*
    2. Principal leaves 6–18 mm broad; lemmas dull, scabrous between the nerves; pedicels thickened upward, less than one-fourth the length of the spikelets (except for the terminal ones).
        3. Principal blades 6–12 mm broad; lemmas obscurely nerved, scabrous, 3.5–5.5 mm long; anthers over 1 mm long_____
        _____2. *G. septentrionalis*
        3. Principal blades 10–18 mm broad; lemmas sharply nerved, hirtellous, 2.5–3.0 mm long; anthers less than 1 mm long____
        _____3. *G. arkansana*

1. Spikelets 2–8 mm long; sheaths terete or subterete.
  4. Lemmas 3–4 mm long, obscurely nerved; spikelets 3–4 mm broad_____4. *G. canadensis*
  4. Lemmas 1.5–2.7 mm long, sharply nerved; spikelets 2.0–2.5 mm broad.
    5. Inflorescence 5–20 cm long; spikelets 2.0–4.5 mm long; first glume 0.5–1.0 mm long; second glume 0.8–1.3 mm long; lemmas 1.5–2.0 mm long_____5. *G. striata*
    5. Inflorescence 20–40 cm long; spikelets 5–6 mm long; first glume 1.2–2.0 mm long; second glume 1.5–2.5 mm long; lemmas 2.0–2.7 mm long_____6. *G. grandis*

1. **Glyceria borealis** (Nash) Batchelder, Proc. Manchester Inst. 1:74. 1900. *Fig. 231.*

Panicularia borealis Nash, Bull. Torrey Club 24:348. 1897.
Perennial, rooting at the lower nodes, with culms to 1.2 m tall; sheaths compressed, glabrous; blades glabrous, 2–6 mm broad; inflorescence 15–45 cm long, ascending; spikelets 10–18 mm long, 6- to 12-flowered, on slender pedicels one-fourth to two-thirds as long; glumes elliptic, obtuse to subacute, obscurely nerved, with a scarious margin, the first 1–2 mm long, the second 2–3 mm long; lemmas obtuse, erose, and scarious at the apex, scabrous only on the nerves, 7-nerved, 3–4 mm long, longer than the palea; anthers less than 1 mm long; 2n = 20 (Church, 1949).

COMMON NAME: Northern Manna Grass.
HABITAT: Shallow water.
RANGE: Newfoundland to Alaska, south to California, Illinois, and New Jersey.
ILLINOIS DISTRIBUTION: Rare; known only from three counties in the northernmost tier of counties. I have not been able to verify the report (Mosher, 1918) of this species from Stark County (*V. H. Chase 100*).
The Illinois collections were made in June when the plants were flowering.
The combination of spikelets over 10 mm long and blades less than 6 mm broad distinguishes this species.

231. *Glyceria borealis* (Northern Manna Grass). *a.* Inflorescence, X½. *b.* Sheath, with ligule, X2½. *c.* Spikelet, X7½. *d.* First glume, X10. *e.* Second glume, X10. *f.* Lemma, X10.

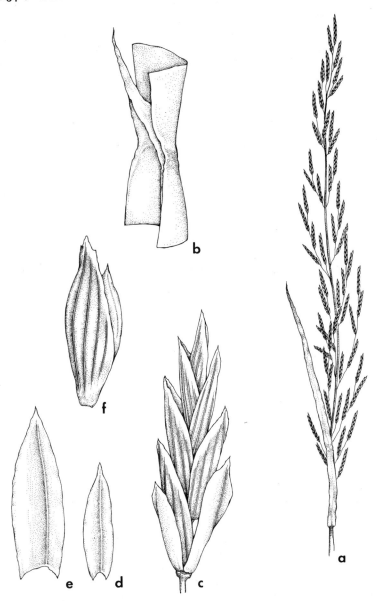

232. *Glyceria septentrionalis* (Manna Grass). *a.* Inflorescence, X½. *b.* Sheath, with ligule, X5. *c.* Spikelet, X5. *d.* First glume, X6. *e.* Second glume, X6. *f.* Lemma and palea, X6.

2. **Glyceria septentrionalis** Hitchc. Rhodora 8:211. 1906. *Fig. 232.*

Panicularia septentrionalis (Hitchc.) Bickn. Bull. Torrey Club 35:196. 1908.

Perennial, rooting at the lower nodes, with culms to 1.5 m tall; sheaths compressed, glabrous; blades glabrous, the principle ones 4–12 mm broad; inflorescence 20–45 cm long, ascending; spikelets 10–20 mm long, 6- to 15-flowered, on upwardly thickened pedicels less than one-fourth as long (except in the terminal spikelets); glumes elliptic to obovate, scarious throughout, obscurely nerved, the first 2–4 mm long, the second 3–5 mm long; lemmas elliptic, obtuse and erose at the apex, obscurely 7-nerved, scabrous between the nerves, 3.5–5.5 mm long slightly shorter than the palea; anthers over 1 mm long; 2n = 40 (Church, 1949).

COMMON NAME: Manna Grass.

HABITAT: Shallow water, marshy soil, swamp meadows.

RANGE: Quebec to Minnesota, south to Texas and Georgia.

ILLINOIS DISTRIBUTION: Occasional; scattered throughout the state.

This species flowers from May to August. It is the most frequent of the species with long spikelets. Until this species was described in 1906, it was known as *G. fluitans* (L.) R. Br., a species far to the north of Illinois.

3. **Glyceria arkansana** Fern. Rhodora 31:49. 1929. *Fig. 233.*

Glyceria septentrionalis var. *arkansana* (Fern.) Steyerm. & Kucera, Rhodora 63:24. 1961.

Perennial, rooting at the lower nodes, with culms to nearly 2 m tall; sheaths compressed, glabrous; blades glabrous, 10–18 mm broad; inflorescence 40–70 cm long, ascending; spikelets 15–20 mm long, 10- to 15-flowered; glumes elliptic to obovate, scarious throughout, rather obscurely nerved, the first 1.5–3.0 mm long, the second 2.5–3.5 mm long; lemmas elliptic, obtuse and erose at the apex, sharply 7-nerved, hirtellous throughout on the back, 2.5–3.0 mm long, slightly shorter than the palea; anthers less than 1 mm long.

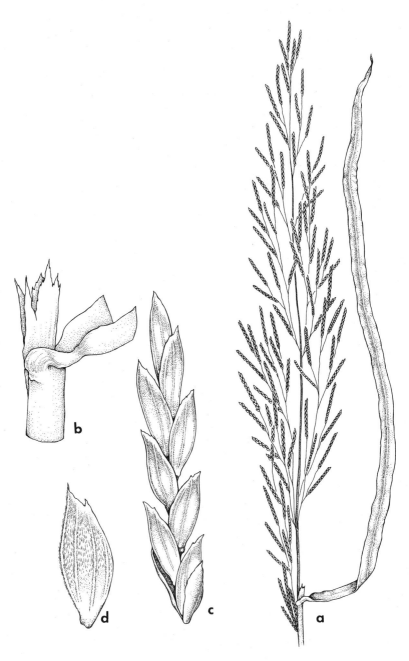

233. *Glyceria arkansana* (Manna Grass). *a.* Inflorescence, X½. *b.* Sheath, with ligule, X2½. *c.* Spikelet, X5. *d.* Lemma, X6.

COMMON NAME: Manna Grass.

HABITAT: Shallow water of swamps.

RANGE: Virginia to Illinois, south to Arkansas and Louisiana; New York (?).

ILLINOIS DISTRIBUTION: Rare; known only from Union County (LaRue Swamp, May 19, 1940, *B. Bauer 2608*, and several subsequent collections from the same place).

This species has been collected in flower in Illinois during May and June.

Steyermark and Kucera believe this taxon to be a southern variety of *G. septentrionalis*, from which it differs by its broader blades, its shorter, more strongly nerved, hirtellous lemmas, and its tiny anthers.

4. **Glyceria canadensis** (Michx.) Trin. Mem. Acad. St. Petersb. VI. Math. Phys. Nat. 1:366. 1830. *Fig. 234.*

Briza canadensis Michx. Fl. Bor. Am. 1:71. 1803.

Panicularia canadensis (Michx.) Kuntze, Rev. Gen. Pl. 2:783. 1891.

Solitary or tufted perennial to 1.5 m tall; sheaths terete or subterete, glabrous; blades scabrous, 2.5–8.5 mm broad; inflorescence open, drooping at the tip, 5–25 cm long; spikelets 4–8 mm long, 3–4 mm broad, 4- to 10-flowered; glumes obscurely nerved, with a scarious margin, the first lanceolate, 1.5–2.5 mm long, the second ovate, 2–3 mm long; lemmas broadly ovate, with a scarious margin, obscurely 7-nerved, 3–4 mm long, longer than the palea; $2n = 60$ (Church, 1949).

COMMON NAME: Rattlesnake Manna Grass.

HABITAT: Wet ground.

RANGE: Newfoundland to Minnesota, south to Illinois and Virginia.

ILLINOIS DISTRIBUTION: Rare; known only from two northern counties. The Pepoon collections cited by Mosher (1918) from Fulton and JoDaviess counties were not located.

This species flowers from late June to mid-September.

5. **Glyceria striata** (Lam.) Hitchc. Proc. Biol. Soc. Wash. 41: 157. 1928.

Poa striata Lam. Tabl. Encycl. 1:183. 1791.

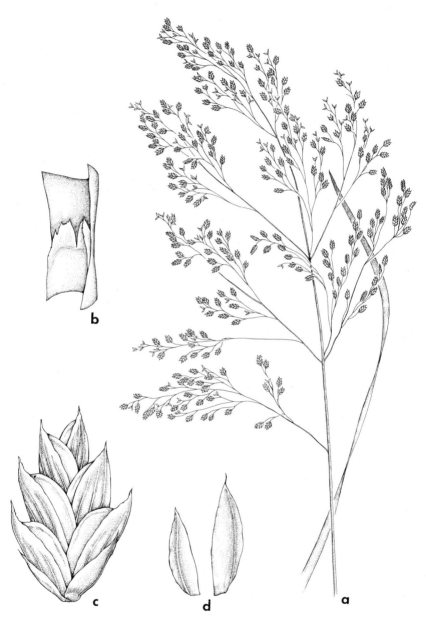

234. *Glyceria canadensis* (Rattlesnake Manna Grass). *a*. Inflorescence, X½. *b*. Sheath, with ligule, X4. *c*. Spikelet, X10. *d*. Glumes, X12½.

Tufted perennial to 1.2 m tall; sheaths terete or subterete, glabrous; blades flat or conduplicate, scabrous above, 2–8 mm broad; inflorescence 5–20 cm long, open, usually drooping at the tip; spikelets 2.0–4.5 mm long, 2.0–2.5 mm broad, 3- to 7-flowered, green or purple; glumes obovate, obscurely nerved, the first 0.5–1.0 mm long, the second 0.8–1.3 mm long; lemmas elliptic to obovate, obtuse, more or less scarious at the apex, 7-nerved, 1.5–2.0 mm long; 2n = 20 (Church, 1949).

Two varieties may be distinguished in Illinois.

1. Spikelets green; uppermost branches of the panicle more or less nodding; lemmas with a minutely scarious apex_____
   _____5a. *G. striata* var. *striata*
1. Spikelets purple; uppermost branches of the panicle ascending; lemmas with a broadly scarious apex_____5b. *G. striata* var. *stricta*

5a. Glyceria striata (Lam.) Hitchc. var. **striata** *Fig. 235.*

Poa nervata Willd. Sp. Pl. 1:389. 1797.
Poa lineata Pers. Syn. Pl. 1:89. 1805.
Glyceria nervata (Willd.) Trin. Mem. Acad. St. Petersb. VI. Math. Phys. Nat. 1:365. 1830.
Panicularia nervata (Willd.) Kuntze, Rev. Gen. Pl. 2:783. 1891.

Leaves flat; inflorescence 10–20 cm long, the uppermost branches more or less nodding; spikelets green, 2–4 mm long; lemmas with a minutely scarious apex.

COMMON NAME: Fowl Manna Grass.
HABITAT: Moist soil.
RANGE: Newfoundland to Alberta, south to Texas and Florida.
ILLINOIS DISTRIBUTION: Common throughout the state; in every county.
This common grass flowers from late May to mid-August. It is distinguished from var. *stricta* primarily by its green spikelets and more open inflorescence.

5b. Glyceria striata (Lam.) Hitchc. var. **stricta** (Scribn.)

Fern. Rhodora 31:47. 1929. *Fig. 236.*

Panicularia nervata stricta Scribn. Bull. U.S.D.A. Div. Agrost. 13:44. 1898.

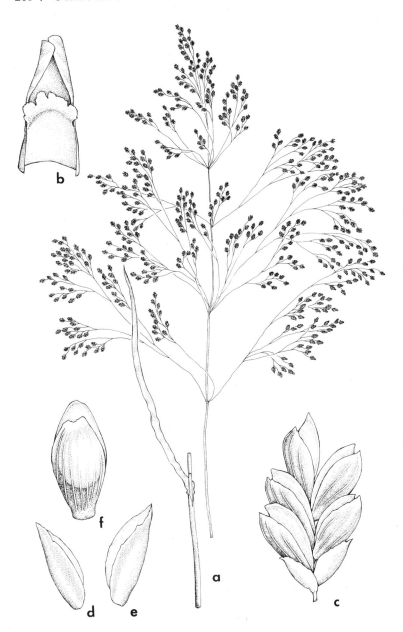

235. *Glyceria striata* var. *striata* (Fowl Manna Grass). *a.* Inflorescence, X½. *b.* Sheath, with ligule, X5. *c.* Spikelet, X12½. *d.* First glume, X25. *e.* Second glume, X25. *f.* Lemma and palea, X25.

Glyceria nervata var. *stricta* Scribn. ex Hitchc. in Gray, Man., ed. 7, 159. 1908.

Leaves flat or plicate; inflorescence 5–15 cm long, the uppermost

236. Glyceria striata var. *stricta* (Fowl Manna Grass). *a.* Inflorescence, X½. *b.* Sheath, with ligule, X5. *c.* Spikelet, X15. *d.* First glume, X17½. *e.* Second glume, X17½.

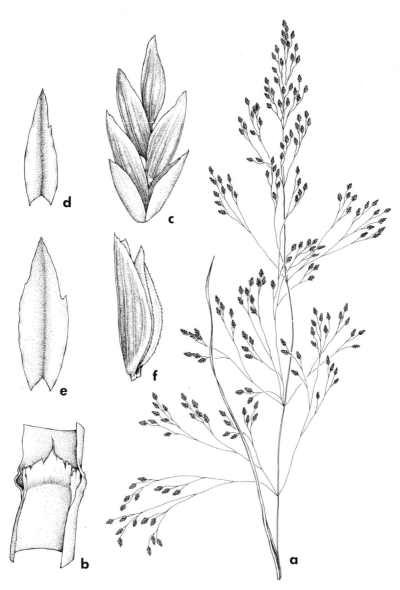

237. *Glyceria grandis* (American Manna Grass). *a.* Inflorescence, X½. *b.* Sheath, with ligule, X5. *c.* Spikelet, X10. *d.* First glume, X15. *e.* Second glume, X15. *f.* Lemma and palea, X15.

branches ascending; spikelets purple, 3.0–4.5 mm long; lemmas with a broadly scarious apex.

COMMON NAME: Fowl Manna Grass.

HABITAT: Wet ground.

RANGE: Labrador to Alaska, south to Mexico, northern Illinois, and New York.

ILLINOIS DISTRIBUTION: Occasional; restricted to the northern one-third of the state.

There is considerable overlapping of characters between this variety and var. *striata* so that it is questionable whether the varieties should be distinguished. Spikelet color seems to be the most reliable character.

6. **Glyceria grandis** S. Wats. ex Gray, Man., ed. 6, 667. 1890.
   Fig. 237.

Poa aquatica var. *americana* Torr. Fl. N. & Mid. U. S. 1:108. 1824.

Glyceria americana (Torr.) Pammel, Rep. Iowa Geol. Surv. 1903:271. 1905.

Panicularia grandis (Gray) Nash in Britt. & Brown, Ill. Fl. 1:265. 1913.

Tufted perennial to 1.5 m tall; sheaths terete or subterete, glabrous; blades glabrous or scabrous, 6–14 mm broad; inflorescence 20–40 cm long, open, nodding at the tip; spikelets 5–6 mm long, 2.0–2.5 mm broad, 2- to 9-flowered, purplish; glumes scarious, acute, the first 1.2–2.0 mm long, the second 1.5–2.5 mm long; lemmas narrowly ovate, obtuse, 7-nerved, 2.0–2.7 mm long; 2n = 20 (Church, 1949).

COMMON NAME: American Manna Grass; Reed Manna Grass.

HABITAT: Newfoundland to Alaska, south to New Mexico, Illinois, and Virginia.

ILLINOIS DISTRIBUTION: Not common; restricted to the extreme northern counties. First collected from Warren, JoDaviess County, by Umbach in July 1896.

This rare species flowers from June to August. The purple spikelets recall *G. striata* var. *stricta,* but *G. grandis* has a more open panicle and longer glumes and lemmas.

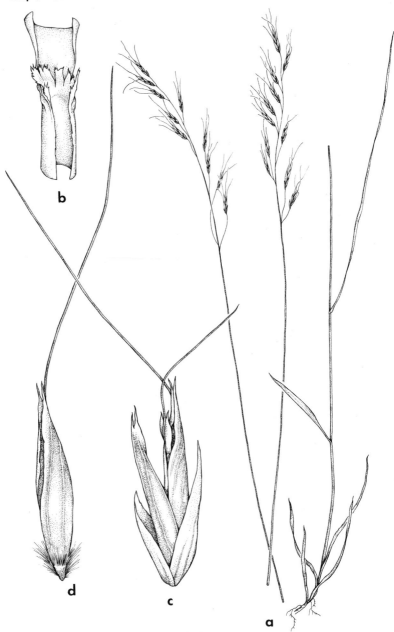

238. *Schizachne purpurascens* (False Melic Grass). *a.* Habit, X½. *b.* Sheath, with ligule, X2½. *c.* Spikelet, X5. *d.* Lemma, X5.

36. *Schizachne* HACK. – False Melic Grass

Perennials; sheaths closed; blades flat; inflorescence paniculate; spikelets 3- to 5-flowered, disarticulating above the glumes; glumes 2, unequal, strongly nerved, shorter than the spikelets; lemmas rounded on the back, prominently nerved, awned, the uppermost sterile.

This genus is related to *Melica* by the presence of sterile or staminate lemmas above the fertile ones. Its lemmas, with a bearded callus, separate it from *Melica*.

Only the following species comprises the genus in Illinois.

1. **Schizachne purpurascens** (Torr.) Swallen, Journ. Wash. Acad. Sci. 18:204. 1928. *Fig. 238.*

Trisetum purpurascens Torr. Fl. N. & Mid. U.S. 1:127. 1824. Loosely tufted perennial, decumbent at the base, with culms to 1 m tall; blades 1–5 mm broad; inflorescence 5–15 cm long, with each drooping branch bearing 1–3 spikelets; spikelets 15–25 mm long, 3- to 5-flowered, more or less purplish; glumes membranous, the first 5.0–6.5 mm long, the second 6–8 mm long; lemmas 5- to 7-nerved, bearded on the callus, the fertile 8–10 mm long, the sterile much smaller; awns 8–15 mm long, divergent at maturity.

COMMON NAME: False Melic Grass.

HABITAT: Moist woodlands.

RANGE: Newfoundland to Alaska, south to New Mexico, Illinois, and Pennsylvania.

ILLINOIS DISTRIBUTION: Rare; known only from JoDaviess County (moist wooded slope, Apple River Canyon, near Stockton, June 16, 1937, *F. J. Hermann 8829*).

Tribe *Stipeae*

Cespitose perennials; inflorescence paniculate; spikelets 1-flowered, disarticulating above the glumes; glumes awnless; lemma obscurely nerved, awned, the awn usually twisted.

This small tribe is represented in Illinois by *Stipa* and *Oryzopsis*.

37. *Stipa* L. – Needle Grass

Perennials; blades flat or usually involute; inflorescence paniculate, open or contracted; spikelets 1-flowered, disarticulating above the glumes; glumes subequal, papery, tapering to a long, slender point; lemma indurate, obscurely nerved, pubescent (at least below), the margins inrolled around the palea, awned.

Species of the western United States are valuable for forage. Hitchcock (1925) has studied the North American species of *Stipa*.

KEY TO THE SPECIES OF Stipa IN ILLINOIS

1. Sheaths villous on the margins and at the summit; ligule less than 1 mm long; glumes 5–11 mm long; lemma 4.5–6.0 mm long, pubescent throughout, the awn 2–4 cm long_____1. *S. viridula*
1. Sheaths more or less glabrous; ligule (at least of the upper leaves) 3–6 mm long; glumes 15–40 mm long; lemma 9–25 mm long, pubescent at base, becoming glabrate above, the awn 10–20 cm long.
    2. Glumes 15–28 mm long; lemma 9–13 mm long, the flexuous but obscurely geniculate awn 10–15 cm long; ligule of upper leaves 3–4 mm long_____2. *S. comata*
    2. Glumes 28–42 mm long; lemma 16–25 mm long, the twice geniculate awn 12–20 cm long; ligule of upper leaves 4–6 mm long_____3. *S. spartea*

1. **Stipa viridula** Trin. Mem. Acad. St. Petersb. Vi. Sci. Nat. 2 (1):39. 1836. *Fig. 239.*

Loosely cespitose perennial with culms to nearly 1 m tall; sheaths villous on the margins and at the summit; ligule less than 1 mm long; blades usually involute, 1–4 mm broad, scaberulous; panicle narrow, contracted, to 20 cm long, the branches ascending; glumes 5–11 mm long, tapering to a long point, the first 3-nerved, the second 5-nerved; lemma fusiform, pale brown, appressed-pubescent throughout, 4.5–6.0 mm long, the awn twice geniculate, 2–4 cm long; $2n = 82$ (Johnson & Rogler, 1943).

COMMON NAME: Feather Grass.

HABITAT: Edge of woods and along railroad track near pond (in Illinois).

RANGE: Minnesota to British Columbia, south to New Mexico and Illinois.

ILLINOIS DISTRIBUTION: Rare; known only from two extreme northern counties. First collected in 1916 by Benke northwest of Pingree Grove, Kane County.

This western species has its easternmost natural stations in Illinois, where it flowers during June and July. It is

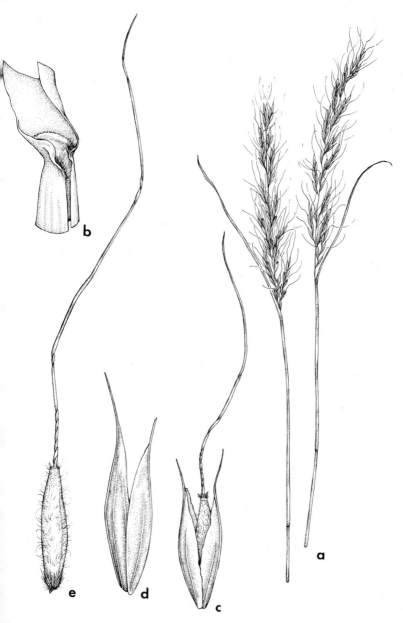

239. *Stipa viridula* (Feather Grass). *a*. Inflorescences, X½. *b*. Sheath, with ligule, X5. *c*. Spikelet, X5. *d*. Glumes, X7½. *e*. Lemma, X7½.

240. Stipa comata (Needle Grass). *a.* Inflorescence, X½. *b.* Sheath, with ligule, X5. *c.* Glumes, X2½. *d.* Lemma, X2½.

very distinct from the other species of *Stipa* in Illinois by its villous sheaths and short ligules, glumes, and lemmas.

2. Stipa comata Trin. & Rupr. Mem. Acad. St. Petersb. VI. Sci. Nat 5(1):75. 1842. *Fig. 240.*

Cespitose perennial with culms to 60 cm tall; sheaths glabrous or nearly so; ligule of upper leaves 3–4 mm long; blades 1–4 mm broad, the basal involute, the upper flat or involute, smooth or scaberulous; panicle narrow, contracted, to 25 cm long, the branches ascending; glumes 15–28 mm long, tapering to a long point, the first 3-nerved, the second 5-nerved; lemma 9–13 mm long, fusiform, pale brown, villous at base, becoming glabrate above, the awn flexuous, obscurely geniculate, 10–15 cm long, pubescent or scabrous; $2n = 44$ (Stebbins & Löve, 1941).

COMMON NAME: Needle Grass.

HABITAT: Dry soil, usually in prairies; loamy soil along railroads.

RANGE: Michigan to Yukon, south to California, Texas, and Indiana.

ILLINOIS DISTRIBUTION: Rare; known only from Winnebago and Cook counties.

This species is closely related to *S. spartea,* but differs in its smaller glumes and lemmas and in its somewhat shorter awns.

It flowers during July and August.

3. Stipa spartea Trin. Mem. Acad. St. Petersb. VI. Math. Phys. Nat. 1:82. 1830. *Fig. 241.*

Tufted perennial with rather stout culms to 1.2 m tall; sheaths glabrous; ligule of upper leaves 4–6 mm long; blades 2–5 mm broad, flat or involute when dry, scabrous and usually pubescent above, glabrous beneath; panicle narrow, to 25 cm long, the branches ascending or slightly nodding; glumes 28–42 mm long, tapering to a point, the first 3-nerved, the second 5-nerved; lemma fusiform, 16–25 mm long, brown, pubescent at the base, the awn twice geniculate, 12–20 cm long, pubescent or scabrous.

241. *Stipa spartea* (Porcupine Grass). *a.* Upper part of plant, X½. *b.* Sheath, with ligule, X2½. *c.* Spikelet, x1.

COMMON NAME: Porcupine Grass.

HABITAT: Sandy soil, particularly in prairies.

RANGE: Ontario to British Columbia, south to New Mexico, Illinois, and Pennsylvania.

ILLINOIS DISTRIBUTION: Rather common in the northern half of the state, but nearly absent in the southern third. This species has the largest features of any *Stipa* in Illinois.

It is also the only species which is found with any regularity. It flowers during May and June.

38. *Oryzopsis* MICHX. –Rice Grass

Slender, tufted perennials; blades flat or involute; inflorescence racemose or paniculate,. contracted or more or less open, few-flowered; spikelets 1-flowered, disarticulating above the glumes; glumes subequal, broad; lemma indurate, broad, obscurely nerved, awned, the margins inrolled partly around the palea; palea indurate.

For an account of the genus *Oryzopsis*, see Johnson (1945).

KEY TO THE SPECIES OF Oryzopsis IN ILLINOIS

1. Blades flat, 5–15 mm broad; spikelets (excluding the awn) 6–9 mm long; glumes acute to acuminate, 7–9 mm long, conspicuously 7-nerved; lemma 5.5–8.5 mm long, with the awn 5–25 mm long.
    2. Upper leaves longer than lower leaves; lemma dark brown to blackish, the awn 12–25 mm long_____1. *O. racemosa*
    2. Upper leaves shorter than lower leaves; lemma pale green to yellowish, the awn 5–10 mm long_____2. *O. asperifolia*
1. Blades involute, 1–2 mm broad; spikelets (excluding the awn) 3–4 mm long; glumes obtuse, 3.5–4.0 mm long, obscurely 5-nerved; lemma 3.5–4.0 mm long, with the awn 1–2 mm long_____
    _____3. *O. pungens*

1. **Oryzopsis racemosa** (J. E. Smith) Ricker in Hitchc. Rhodora 8:210. 1906. *Fig. 242.*

Milium racemosum J. E. Smith in Rees, Cycl. 23:Milium no. 15. 1813.

Oryzopsis melanocarpa Muhl. Descr. Gram. 79. 1817.

Cespitose perennial from rhizomes; culms to 1 m tall; upper leaves longer than lower leaves, 5–15 mm broad, short-pilose above, scabrous beneath; panicle contracted, sparsely branched, to 25 cm long, the branches spreading to ascending; spikelets (excluding the awn) 7–9 mm long; glumes narrowly elliptic, acute to acuminate, 7–9 mm long, 7-nerved; lemma dark brown to blackish, shining, pubescent, 5.5–8.0 mm long, with the awn 12–25 mm long; 2n = 46 (Johnson, 1945).

COMMON NAME: Rice Grass.
HABITAT: Rich, rocky woodlands.
RANGE: Quebec to North Dakota, south to Missouri and Virginia.
ILLINOIS DISTRIBUTION: Rare; known from four north-central counties.
This rare species has been collected in Illinois in July and September. The color of the lemma and the long awn distinguish it from *O. asperifolia*.

I have not seen the specimen collected by Welsch from St. Clair County nor the specimen collected by Johnson from Cook County, as reported by Mosher (1918).

2. **Oryzopsis asperifolia** Michx. Fl. Bor. Am. 1:51. 1803. *Fig. 243.*

Cespitose perennial; culms slender to stoutish, spreading or sometimes prostrate, to 70 cm tall; basal leaves longer than the much reduced upper leaves, 5–10 mm broad, pubescent and glaucous above, scabrous beneath; raceme contracted, slender, to 12 cm long; spikelets (excluding the awn) 6–8 mm long; glumes elliptic, acute or mucronate at the short-ciliate apex, 7.0–8.5 mm long, 7-nerved; lemma pale green or yellowish, pubescent, 7.0–8.5 mm long, the awn 5–10 mm long; 2n = 46 (Johnson, 1945).

COMMON NAME: Rice Grass.
HABITAT: Rather dry woodlands.
RANGE: Newfoundland to British Columbia, south to New Mexico, Illinois, and West Virginia.
ILLINOIS DISTRIBUTION: Very rare; collected only a single time, from Cook County, in 1877 (specimen in the herbarium of Southern Illinois University).
This species blooms earlier than *O. racemosa*, coming into flower from late April to early July.

3. **Oryzopsis pungens** (Torr.) Hitchc. Contr. U. S. Nat. Herb. 12:151. 1908. *Fig. 244.*

242. *Oryzopsis racemosa* (Rice Grass). *a.* Upper part of plant, X½. *b.* Sheath, with ligule, X5. *c.* Spikelet, X6. *d.* Lemma, X6.

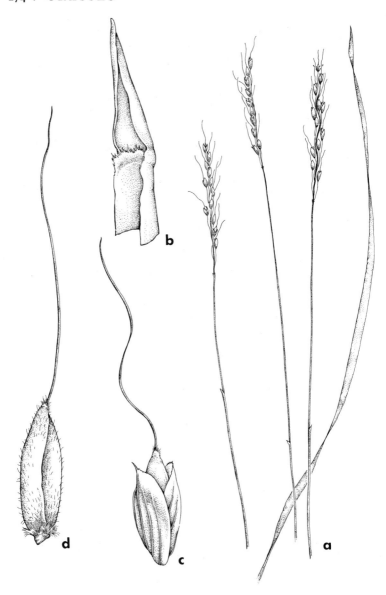

243. Oryzopsis asperifolia (Rice Grass). *a.* Inflorescences, X½. *b.* Sheath, with ligule, X5. *c.* Spikelet, X5. *d.* Lemma, X7.

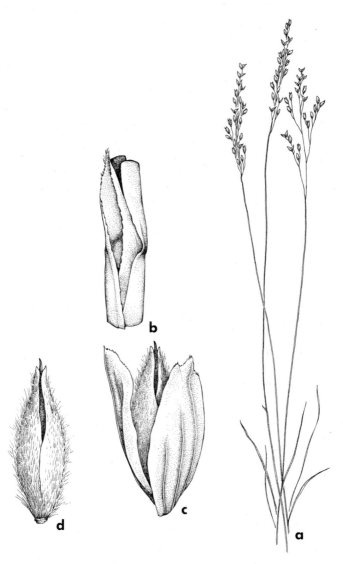

244. Oryzopsis pungens (Rice Grass). *a.* Inflorescences, X½. *b.* Sheath, with ligule, X5. *c.* Spikelet, X12½. *d.* Lemma, X12½.

Milium pungens Torr. in Spreng. Neu. Entd. 2:102. 1821.
Densely tufted perennial; culms slender, erect, to 50 cm tall;
sheaths more or less scabrous; blades involute when dry, 1–2 mm
broad, scabrous, the upper leaves much reduced; panicle slender,
contracted at first, later open, to 8 cm long; spikelets (excluding
the awn) 3.5–4.0 mm long; glumes elliptic-obovate, obtuse, in-
conspicuously 5-nerved, 3.5–4.0 mm long; lemma gray or pale
green, appressed-pubescent, 3.5–4.0 mm long, the awn 1–2 mm
long; 2n = 22 (Johnson, 1945).

COMMON NAME: Rice Grass.
HABITAT: Dry soil.
RANGE: Quebec to British Columbia, south to Colorado,
Illinois, and Pennsylvania.
ILLINOIS DISTRIBUTION: Very rare; collected only from
Menard County by E. Hall in the nineteenth century.
The report from St. Clair County by Mosher (1918)
based on a specimen collected by Welsch could not be
verified.
This species flowers from late April to late June. It is
very distinct from the other species of *Oryzopsis*.

Tribe *Brachyelytreae*

Rhizomatous perennial; inflorescence paniculate; spikelets 1-flow-
ered, disarticulating above the glumes; glumes minute, the lower
sometimes absent; lemma 5-nerved, awned.
    Only the genus *Brachyelytrum* represents this tribe in Illinois.

39. *Brachyelytrum* BEAUV.

Rhizomatous perennial; blades flat; panicle contracted, few-flowered; spikelets 1-flowered, disarticulating above the glumes;
glumes minute, unequal; lemma rounded on the back, 5-nerved,
awned; palea nearly as long as the lemma, 2-keeled.
    Only the following species comprises the genus.

1. **Brachyelytrum erectum** (Schreb.) Beauv. Ess. Agrost. 155.
    1812. *Fig. 245.*
Muhlenbergia erecta Schreb. in Roth, Neue Beytrage Bot.
1:97. 1802.
Dilepyrum aristosum Michx. Fl. Bor. Am. 1:40. 1803.
Muhlenbergia aristata Pers. Syn. Pl. 1:73. 1805.

245. Brachyelytrum erectum. a. Upper part of plant, X½. *b.* Sheath, with ligule, X5. *c.* Spikelet, X5. *d.* Lemma, X5.

Brachyelytrum aristatum (Pers.) Roem. & Schult. Syst. Veg. 2:413. 1817.

Perennial from short, knotty rhizomes; culms erect, to nearly 1 m tall, glabrous or puberulent; sheaths sparsely retrorsely pubescent; blades scabrous, sparsely pubescent beneath, to 15 mm broad; panicle narrow, contracted, erect to slightly nodding, to 15 cm long; glumes 1-nerved, glabrous, the first absent or less than 1 mm long, the second subulate, 1–4 mm long; lemma linear-subulate, 5-nerved, hispidulous on the nerves, 6–10 mm long, with the awn 10–25 mm long; 2n = 22 (Brown, 1950).

HABITAT: Moist or occasionally dry woodlands.

RANGE: Newfoundland to Ontario, south to Oklahoma and Georgia.

ILLINOIS DISTRIBUTION: Occasional throughout the state. This species is sometimes mistaken for immature specimens of *Bromus*. The mature spikelet has but a single flower. It flowers from late May to early August.

Tribe *Diarrheneae*

Only the genus *Diarrhena*, usually placed in the Festuceae, comprises this tribe.

40. Diarrhena BEAUV.

Erect perennials; leaves flat; inflorescence paniculate; spikelets 3- to 5-flowered, disarticulating above the glumes; glumes unequal, shorter than the lemmas; lemmas 3-nerved, awnless; stamens 1–3; grain large, exserted from the floret.

Of the five species of *Diarrhena*, only a variety of one of them occurs in Illinois.

1. **Diarrhena americana** Beauv. var. **obovata** Gleason, Phytologia 4:21. 1952. *Fig. 246.*

Festuca diandra Michx. Fl. Bor. Amer. 1:67. 1803, non Moench. (1794).

Diarina festucoides Raf. Med. Repos. N. Y. 5:352. 1808.

Diarrhena diandra (Michx.) Wood, Class-book, ed. 2, 612. 1847.

Corycarpus diandrus (Michx.) Kuntze, Rev. Gen. Pl. 2:772. 1891.

246. *Diarrhena americana* var. *obovata.* *a.* Inflorescences, X½. *b.* Sheath, with ligule, X5. *c.* Spikelet, X5. *d.* First glume, X6. *e.* Second glume, X6. *f.* Grain, X3½.

Slender perennials from creeping rhizomes; culms glabrous, to nearly 1 m tall; leaves broad, flat, glabrous, nearly as long as the culm, to 1.5 (−1.8) cm broad; panicle sparsely branched, scabrous, to 30 cm long; spikelets to 15 mm long, 3- to 5-flowered; glumes unequal, the first to 3 mm long, 1-nerved, the second to

4.5 mm long, 3- to 5-nerved; lemmas firm, glabrous, mucronate, to 7 mm long, mostly 3-nerved, the upper sterile; palea firm, 2-nerved; grain obtusely beaked, exserted from the spreading lemma and palea, 5–6 mm long.

HABITAT: Low, shaded woods; moist ledges; base of limestone cliffs.

RANGE: West Virginia to South Dakota, south to Texas and Georgia.

ILLINOIS DISTRIBUTION: Throughout the state, except for the northeastern counties. Several reports of this plant from Cook County have not been verified.

This is one of the more handsome woodland grasses in Illinois.

The spikelets mature during midsummer. The bottle-shaped grain is distinctive.

Typical var. *americana,* with pubescent leaf sheaths, hirsutulous panicle branches, and larger spikelets, apparently does not occur in Illinois, although it is known from Indiana and southwestern Missouri.

SUBFAMILY Panicoideae

Annuals or perennials; leaves various; spikelets with one fertile and one sterile or staminate floret, disarticulating below the glumes.

Under the system of classification followed here, the subfamily Panicoideae is composed of tribes Paniceae and Andropogoneae.

Tribe *Paniceae*

Annuals, or tufted or rhizomatous perennials; inflorescence a panicle or raceme, sometimes digitate; spikelets with 1 perfect flower; first glume frequently minute or absent; lemma of sterile floret similar in texture to second glume.

This tribe is represented in Illinois by nine genera, including *Panicum* which has the most number of species of any genus of grasses in the state. The other genera are *Digitaria, Trichachne, Leptoloma, Eriochloa, Paspalum, Echinochloa, Setaria,* and *Cenchrus.*

41. *Digitaria* HEIST. – Finger Grass

Annuals or perennials; blades flat; inflorescence racemose, digitate; spikelets 1-flowered, solitary or in groups of 2 or 3, alternately disposed in two rows on one side of a 3-angled rachis; first glume minute, sometimes absent; second glume and sterile lemma 5-nerved; fertile lemma cartilaginous, with hyaline margins.

KEY TO THE SPECIES OF Digitaria IN ILLINOIS

1. Culms rooting at the lower nodes, decumbent at the base; rachis broadly winged, about 1 mm broad.
    2. Sheaths (at least the lower) papillose-pilose; blades pilose to scabrous; spikelets 2.5–3.5 mm long; second glume about half as long as spikelet, usually 1.2–1.6 mm long; fertile lemma greenish-brown_____1. _D. sanguinalis_
    2. Sheaths glabrous; blades glabrous; spikelets 1.7–2.2 mm long; second glume about as long as spikelet, 1.7–2.2 mm long; fertile lemma dark brown to blackish_____2. _D. ischaemum_
1. Culms erect, not rooting at the lower nodes; rachis narrowly winged, less than 1 mm broad.
    3. Racemes less than 10 cm long; spikelets 1.5–1.7 (–2.0) mm long; second glume and sterile lemma more or less glabrous to short-pubescent, 1.5–1.7 (–2.0) mm long_____3. _D. filiformis_
    3. Racemes over 10 cm long; spikelets 2.0–2.5 mm long; second glume and sterile lemma long-pubescent, 2.0–2.5 mm long____
    _____4. _D. villosa_

1. **Digitaria sanguinalis** (L.) Scop. Fl. Carn. 1:52. 1772.
Panicum sanguinale L. Sp. Pl. 57. 1753.
Syntherisma sanguinalis (L.) Dulac, Fl. Haut. Pyr. 77. 1867.
Decumbent or prostrate annual; culms much branched, rooting at the nodes, to 75 cm long; sheaths (at least the lower) papillose-pilose; blades 5–10 mm broad, pilose to scabrous; racemes 3–12, in 1–3 whorls, to 20 cm long; rachis broadly winged, about 1 mm broad, scabrous on the margins; spikelets mostly paired, 2.5–3.5 mm long; first glume minute; second glume narrow, ciliate, 1.2–1.6 mm long, 5- to 7-nerved; sterile lemma 2.5–3.5 mm long, 5-nerved, scabrous, appressed-pubescent, or with cilia to 1.5 mm long; fertile lemma acute, minutely pitted, greenish-brown; 2n = 36 (Avdulov, 1931), 36, 48 (Brown, 1948).

Two varieties may be distinguished in Illinois.

1. Spikelets 2.5–3.0 mm long; sterile lemma appressed-pubescent____
    _____1a. _D. sanguinalis_ var. _sanguinalis_
1. Spikelets 3.0–3.5 mm long; sterile lemma with cilia to 1.5 mm long.
    _____1b. _D. sanguinalis_ var. _ciliaris_

1a. Digitaria sanguinalis (L.) Scop. var. sanguinalis *Fig. 247 a–d.*

Spikelets 2.5–3.0 mm long; sterile lemma appresed-pubescent.

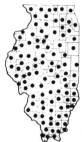

COMMON NAME: Crab Grass.
HABITAT: Waste ground, lawns.
RANGE: Native of Europe and Asia; established throughout the United States.
ILLINOIS DISTRIBUTION: Very common; in every county. This is the common Crab Grass which is so troublesome and ubiquitous in lawns. It flowers from late June to mid-October.

1b. Digitaria sanguinalis (L.) Scop. var. ciliaris (Retz.) Parl. Fl. Ital. 1:126. 1848. *Fig. 247e.*

Panicum ciliare Retz. Obs. Bot. 4:16. 1786.
Digitaria sanguinalis ssp. *ciliaris* (Retz.) Domin, Preslia 13/15:47. 1935.

Spikelets 3.0–3.5 mm long; sterile lemma with cilia to 1.5 mm long.

HABITAT: Waste ground.
RANGE: Native of Europe and Asia; known in the United States from only a few states.
ILLINOIS DISTRIBUTION: Apparently rare; known from Perry County.

2. Digitaria ischaemum (Schreb.) Schreb. ex Muhl. Descr. Gram. 131. 1817. *Fig. 248.*

Panicum ischaemum Schreb. in Schweigger, Spec. Fl. Erland. 16. 1804.
Digitaria humifusa Pers. Syn. Pl. 1:85. 1805.
Syntherisma glabrum Schrad. Fl. Germ. 1:163. 1806.
Panicum glabrum (Schrad.) Gaud. Agrost. Helv. 1:22. 1811.
Digitaria glabra (Schrad.) Beauv. Ess. Agrost. 51. 1812.
Panicum glabrum var. *mississippiense* Gattinger, Tenn. Fl. 95. 1887, name only.

247. *Digitaria sanguinalis* (Crab Grass).—*var. sanguinalis.* *a.* Upper part of plants, X½. *b.* Sheath, with ligule, X4. *c.* Spikelet, front view, X10. *d.* Spikelet, back view, X10.—var. *ciliaris.* *e.* Spikelets, X15.

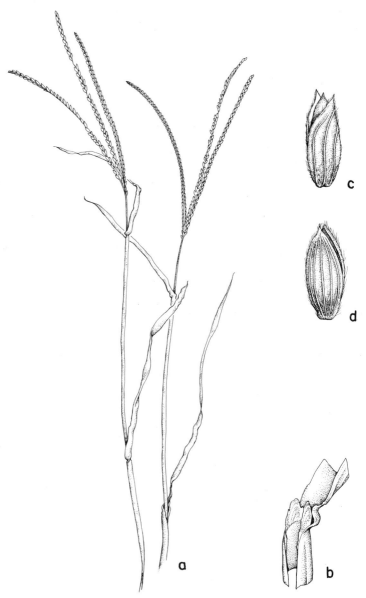

248. *Digitaria ischaemum* (Smooth Crab Grass). *a*. Inflorescences, X½. *b*. Sheath, with ligule, X4. *c*. Spikelet, front view, X12½. *d*. Spikelet, back view, X12½.

Digitaria ischaemum var. *mississippiensis* (Gattinger) Fern.
Rhodora 22:103. 1920.

Decumbent or prostrate annual; culms much branched, rooting
at the nodes, to 80 cm long; sheaths glabrous; blades 3–6 mm
broad, glabrous; racemes 2–7, to 13 cm long, purplish or greenish;
rachis broadly winged, about 1 mm broad, glabrous; spikelets
solitary or paired, 1.7–2.2 mm long; first glume minute or wanting; second glume pubescent, 1.7–2.2 mm long, 5- to 7-nerved;
sterile lemma 1.7–2.2 mm long, pubescent; fertile lemma acute,
minutely pitted, dark brown to blackish; 2n = 36 (Brown, 1948).

COMMON NAME: Smooth Crab Grass.

HABITAT: Waste ground.

RANGE: Native of Europe; introduced in the United
States from Quebec to Washington, south to California
and South Carolina.

ILLINOIS DISTRIBUTION: Not uncommon; throughout the
state.

Some specimens tend to be tinged with purple on various vegetative structures.

This species is smaller than *D. sanguinalis* and has glabrous vegetative structures. More robust specimens with greenish
racemes to 13 cm long have been called var. *mississippiensis*.
There is no character separating var. *ischaemum* from var. *mississippiensis* which does not intergrade hopelessly.

Smooth Crab Grass flowers from mid-July to mid-October.

3. **Digitaria filiformis** (L.) Koel. Descr. Gram. 26. 1802. *Fig. 249.*

Panicum filiforme L. Sp. Pl. 57. 1753.

Syntherisma filiformis (L.) Nash, Bull. Torrey Club 22:420.
1895.

Annual; culms erect or ascending, much branched, to nearly 1 m
tall; upper sheaths glabrous, the lower glabrous to sparsely pilose
to hirsute; blades 1–4 mm broad, scabrous above, hirsute to glabrous below; racemes 2–6, to 10 cm long; rachis narrow, less than
1 mm broad; spikelets paired or in 3s, 1.5–1.7 (–2.0) mm long;
first glume absent; second glume and sterile lemma glabrous to
short-pubescent, 1.5–1.7 (–2.0) mm long; fertile lemma dark
brown or purple, 1.5–1.7 (–2.0) mm long; 2n = 36 (Brown,
1948).

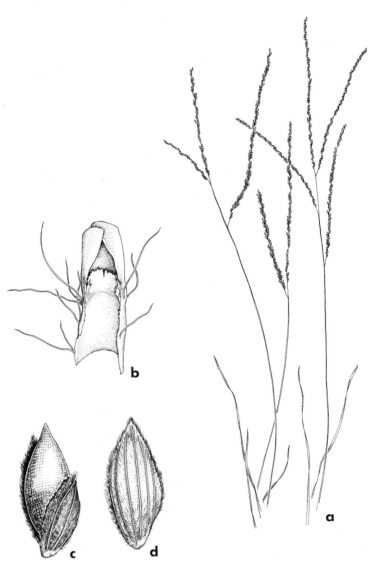

249. *Digitaria filiformis* (Slender Crab Grass). *a.* Inflorescences, X½. *b.* Sheath, with ligule, X5. *c.* Spikelet, front view, X32½. *d.* Spikelet, back view, X32½.

COMMON NAME: Slender Crab Grass; Finger Grass.

HABITAT: Sandy soil.

RANGE: New Hampshire to Iowa, south to Texas and Florida; Mexico.

ILLINOIS DISTRIBUTION: Occasional in central Illinois, absent in the extreme north, and rare in the extreme south. This is the most slender species of *Digitaria* in Illinois. It has the shortest spikelets of any Crab Grass. It does not tend to become weedy. Flowering time is early August to late September.

4. **Digitaria villosa** (Walt.) Pers. Syn. Pl. 1:85. 1805. *Fig. 250.*

Syntherisma villosa Walt. Fl. Carol. 77. 1788.

Digitaria filiformis var. *villosa* (Walt.) Fern. Rhodora 36:19. 1934.

Annual; culms erect or ascending, much branched, sometimes over 1 m tall; upper sheaths glabrous or pilose; lower sheaths densely pilose; blades 3–6 mm broad, softly pilose to rarely nearly glabrous; racemes 2–6, over 10 cm long, sometimes to 25 cm long; rachis narrow, less than 1 mm broad; spikelets paired or in 3s, 2.0–2.5 mm long; first glume absent; second glume and sterile lemma densely pubescent with long hairs, 2.0–2.5 mm long; fertile lemma dark brown, 2.0–2.5 mm long.

COMMON NAME: Hairy Finger Grass.

HABITAT: Sandy soil.

RANGE: Virginia to Kansas, south to Texas and Florida.

ILLINOIS DISTRIBUTION: Rare; known only from Jackson County (Giant City State Park, July 31, 1964, *R. H. Mohlenbrock 13717*).

This species is similar to *D. filiformis* and is considered a variety of it by some authors. *Digitaria villosa*, however, is generally more robust all around, and has larger spikelets and more pubescent glumes and sterile lemmas. It flowers from late July to mid-September.

42. *Trichachne* NEES – Sour Grass

Perennials from swollen bases; blades flat; inflorescence paniculate, composed of ascending racemes; spikelets 1-flowered, short-pedicelled, borne in pairs in two rows along one side of the rachis; first glume minute; second glume and sterile lemma simi-

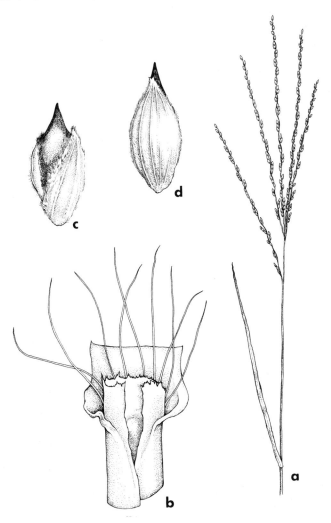

250. *Digitaria villosa* (Hairy Finger Grass). *a.* Inflorescence, X½. *b.* Sheath, with ligule, X7½. *c.* Spikelet, front view, X15. *d.* Spikelet, back view, X15.

lar, 3- to 5-nerved, silky; fertile lemma cartilaginous, acuminate, with a hyaline margin. Some authors would combine this genus with *Digitaria*.

Only the following species occurs in Illinois.

1. **Trichachne insularis** (L.) Nees, Agrost. Bras. 86. 1829. *Fig. 251.*

Andropogon insularis L. Syst. Nat., ed. 10, 2:1304. 1759.
Perennial; culms erect, to 1.5 m tall; sheaths more or less hirsute; blades glabrous or ciliate, to 15 mm broad; panicle to 30 cm long, the racemes to 15 cm long; spikelets 3.5–4.5 mm long (excluding the silky hairs); first glume 0.5 mm long, glabrous; second glume, sterile lemma, and fertile lemma lanceolate, acuminate, 3.5–4.5 mm long, with long, tawny hairs; 2n = 36 (Brown, 1951).

COMMON NAME: Sour Grass.
HABITAT: Along highway, in ditch (in Illinois).
RANGE: Adventive in Illinois; native of southeastern United States; Mexico; West Indies; South America.
ILLINOIS DISTRIBUTION: Known only from Williamson County (along road between Illinois Route 13 and Cambria, *J. W. Voigt* in 1954). The nearest station to Illinois for this grass is in the southeastern United States. It is almost certain that Sour Grass is adventive in Illinois.

Gould (1968) and others propose to place the genus *Trichachne* in *Digitaria,* a view rejected in this work.

43. *Leptoloma* CHASE – Fall Witch Grass

Tufted perennials; blades flat; inflorescence paniculate, terminal, diffuse; spikelets 1-flowered, solitary at the end of long, capillary pedicels; first glume minute or absent; second glume and both lemmas about equal in length, the second glume 3-nerved, the sterile lemma 5- to 7-nerved, the fertile lemma rugulose, cartilaginous.

Leptoloma is distinguished from *Panicum* primarily by its cartilaginous fertile lemma.

There has been a proposal by Henrard (1950) to include *Leptoloma* within *Digitaria.*

Only the following species occurs in Illinois.

1. **Leptoloma cognatum** (Schult.) Chase, Proc. Biol. Soc. Wash. 19:192. 1906. *Fig. 252.*

Panicum divergens Muhl. ex Ell. Bot. S. C. & Ga. 1:130. 1816, non HBK. (1815).
Panicum cognatum Schult. Mantissa 2:235. 1824.
Panicum autumnale Bosc ex Spreng. Syst. Veg. 1:320. 1825.

251. *Trichachne insularis* (Sour Grass). *a.* Inflorescence, X½. *b.* Sheath, with ligule, X7½. *c.* Spikelet, front view, X12½. *d.* Spikelet, back view, X12½.

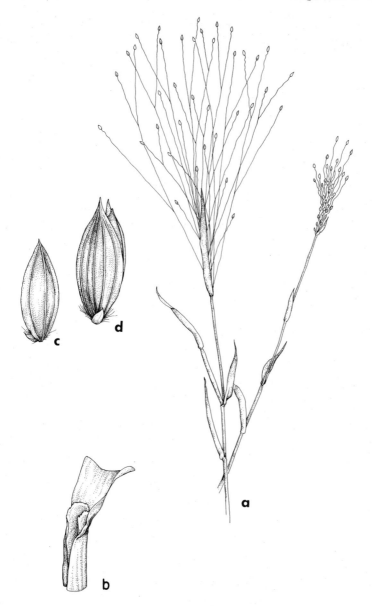

252. *Leptoloma cognatum* (Fall Witch Grass). *a.* Inflorescences, X½. *b.* Sheath, with ligule, X5. *c.* Spikelet, front view, X12½. *d.* Spikelet, back view, X12½.

Digitaria cognata (Schult.) Pilger in Engl. & Prantl, Pflanzenf. 14e:50. 1940.

Densely tufted perennial; culms much branched from the base, to 70 cm tall; upper sheaths more or less glabrous, the lower pilose; blades 4–6 mm broad, scabrous on the margins; panicle 15–35 cm long, spreading, diffusely branched, purplish, the axils pilose; pedicels to 8 cm long, capillary, 3-angled, scabrous; spikelets acute to acuminate, 2.5–3.0 mm long; second glume and sterile lemma with appressed silky pubescence between the nerves and the margins; 2n = 36 (Brown, 1948).

COMMON NAME: Fall Witch Grass.

HABITAT: Sandy soil.

RANGE: New Hampshire to Minnesota, south to Arizona and Florida; Mexico.

ILLINOIS DISTRIBUTION: Occasional throughout the state; apparently more common along the Illinois River.

This species is similar to species of *Digitaria,* except for the long-pedicellate, solitary spikelets. The capillary pedicels remind one of *Eragrostis capillaris* or *Panicum capillare.* This species flowers from July to September.

Fall Witch Grass becomes highly branched at the base, which results in a sprawling habit. During autumn, the panicle becomes purplish and breaks off in its entirety, thus recalling a Tumbleweed.

44. *Eriochloa* HBK. – Cup Grass

Annuals or perennials; inflorescence a contracted panicle; spikelets with one perfect floret and one sterile or staminate floret, borne in two rows on one side of the rachis, disarticulating below the glumes; first glume minute and fused with the rachis node; second glume and sterile lemma subequal, acute; fertile lemma indurate, often short-awned.

The genus is composed primarily of tropical and subtropical species, some of which serve as good forage grasses. All the species of *Eriochloa* in Illinois are adventive.

Eriochloa differs from *Digitaria* by its indurate fertile lemmas and from *Paspalum* by its awned or apiculate fertile lemmas.

KEY TO THE SPECIES OF *Eriochloa* IN ILLINOIS

1. Pedicels and rachis villous; spikelets about 5 mm long _ _1. *E. villosa*
1. Pedicels and rachis short-pilose; spikelets 3.5–5.0 mm long.

2. Grain 2.0–2.5 mm long, with an awn to 1 mm long; blades pubescent, 3–7 mm broad_____2. *E. contracta*
2. Grain 3 mm long, apiculate; blades glabrous, 5–10 mm broad__ _____3. *E. gracilis*

1. Eriochloa villosa (Thunb.) Kunth, Rev. Gram. 1:30. 1829. *Fig. 253.*

Paspalum villosum Thunb. Fl. Jap. 45. 1784.

Annual to 80 cm tall, more or less villosulous; ligule composed of short hairs; blades 3–8 mm broad, villosulous; panicle to 15 cm long, composed of more or less ascending racemes; rachis and pedicels villous; spikelets subacute to obtuse, about 5 mm long; $2n = 54$ (Avdulov, 1928).

COMMON NAME: Cup Grass.

HABITAT: Weed in cornfield.

RANGE: Native of Asia; adventive or introduced in Colorado, Oregon, and Illinois.

ILLINOIS DISTRIBUTION: Known from Livingston County (cornfield, 3 miles east of Odell, August 25, 1950, *R. A. Evers & J. V. Myers 26812, 26813.* Verified by Jason R. Swallen). Very recently collected in Cook County.

This rarely escaped introduction can be readily distinguished from other species of *Eriochloa* in Illinois by its very hairy pedicels and rachises.

2. Eriochloa contracta Hitchc. Proc. Biol. Soc. Wash. 41:163. 1928. *Fig. 254.*

Helopus mollis C. Muell. Bot. Ztg. 19:314. 1861, non *E. mollis* Kunth (1829).

Tufted annual; culms to 75 cm tall, hirsutulous; ligule a fringe of hairs; blades 3–7 mm broad, pubescent; panicle to 15 cm long, composed of up to 25 racemes to 2 cm long; rachis and pedicels pilosulous, purplish, the pedicels to 1 mm long; spikelets lanceoloid or lance-ovoid, 3.5–4.0 mm long; second glume and sterile lemma acuminate to very shortly awned, appressed-pubescent; fertile lemma with an awn to 1 mm long; grain 2.0–2.5 mm long; $2n = 36$ (Brown, 1950).

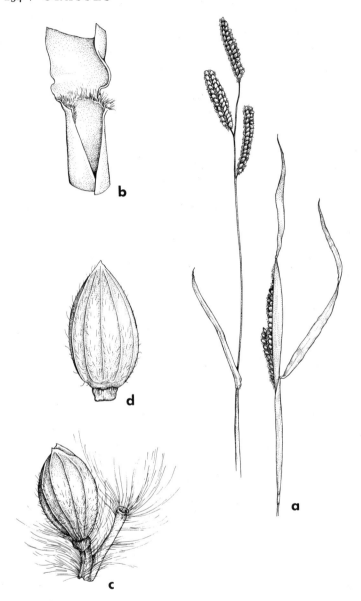

253. *Eriochloa villosa* (Cup Grass). *a.* Inflorescences, X½. *b.* Sheath, with ligule, X5. *c.* Spikelet, with adjacent pedicel, X5. *d.* Spikelet, X7.

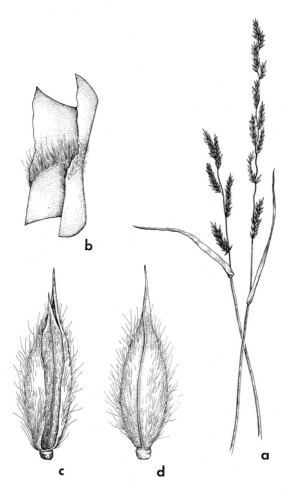

254. Eriochloa contracta (Prairie Cup Grass). *a.* Inflorescences, X½. *b.* Sheath, with ligule, X5. *c.* Spikelet, front view, X12½. *d.* Spikelet, back view, X12½.

COMMON NAME: Prairie Cup Grass.

HABITAT: Moist soil.

RANGE: Native from Missouri westward; adventive in Illinois.

ILLINOIS DISTRIBUTION: Known from Jackson and Union counties, where the first collections in each county were made in 1954.

The hirsululous stems, pubescent blades, short racemes, awned fertile lemma, and short grain distinguish this species from *E. gracilis*.

3. Eriochloa gracilis (Fourn.) Hitchc. Journ. Wash. Acad. Sci. 23:455. 1933. *Fig. 255.*

Helopus gracilis Fourn. Mex. Pl. 2:13. 1886.

Annual; culms to 85 cm tall, more or less glabrous; ligule a fringe of short hairs; blades 5–10 mm broad, glabrous; panicle to 15 cm long, composed of ascending racemes to 4 cm long; rachis and pedicels pilosulous; spikelets acuminate, 4–5 mm long; second glume and sterile lemma acuminate to short-awned, appressed-pubescent; fertile lemma apiculate; grain about 3 mm long.

COMMON NAME: Cup Grass.

HABITAT: Field border.

RANGE: Native of the western United States; adventive in Illinois.

ILLINOIS DISTRIBUTION: Known only from Union County (4 miles northwest of Ware, October 22, 1958, *R. A. Evers 63127*).

This species is generally larger in all respects than *E. contracta*.

45. Paspalum L. – Bead Grass

Annuals or perennials; inflorescence of many 1-flowered, usually plano-convex, subsessile, solitary or paired spikelets arranged along a central axis in 2 or 4 rows with the convex sides toward the rachis, forming simple spike-like racemes; racemes 1-many, digitate or racemose, terminal; first glume usually wanting; second glume similar to sterile lemma; fertile lemma and palea chartaceous-indurate, the margins of the lemma inrolled at maturity; stamens 3; styles 2; stigmas plumose.

For a detailed account of the Illinois species of *Paspalum*, see Verts and Mohlenbrock (1966).

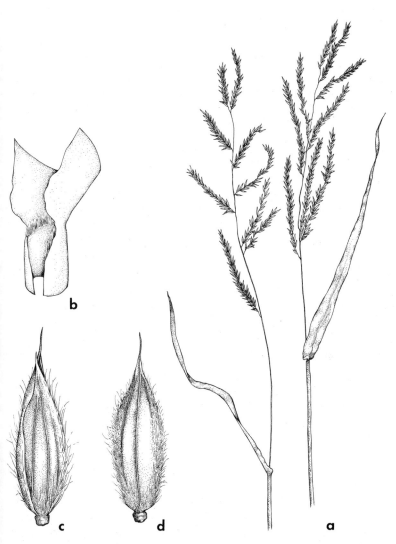

255. *Eriochloa gracilis* (Cup Grass). *a.* Inflorescences, X½. *b.* Sheath, with ligule, X5. *c.* Spikelet, front view, X12½. *d.* Spikelet, back view, X12½.

KEY TO THE TAXA OF Paspalum IN ILLINOIS

1. Rachis foliaceous, the margins folded over and clasping the spikelets or their bases.
   2. Racemes of each inflorescence 1–5; rachis shorter than the rows of spikelets_____1. *P. dissectum*
   2. Racemes of each inflorescence 5–50, usually more than 10; rachis longer than the rows of spikelets_____2. *P. fluitans*
1. Rachis firm, narrow or broad, but the margins not folded over the rows of spikelets.
   3. Rachis broad, over 1.5 mm wide; spikelets arranged in 4 rows___ _____3. *P. pubiflorum* var. *glabrum*
   3. Rachis narrower, less than 1.5 mm wide (about 1.5 mm wide in *P. lentiferum*); spikelets in 2 rows (4 in some racemes of *P. floridanum*).
      4. Spikelets 3.6 mm long or longer; culms robust, 1–2 m tall__ _____4. *P. floridanum*
      4. Spikelets less than 3.2 mm long; culms slender, usually less than 1 m tall (occasionally to 1.5 m in *P. lentiferum*).
         5. Spikelets 2.5–3.2 (–3.4) mm long; sterile lemma 5-nerved, with lateral nerves approximate at the margins.
            6. Spikelets solitary; leaves glabrous to sparsely pilose__ _____5. *P. laeve*
            6. Spikelets paired, or paired and solitary in the same raceme; leaves pilose, becoming villous at base_____ _____6. *P. lentiferum*
         5. Spikelets 1.8–2.4 mm long; sterile lemma 3-nerved, the marginal nerves obscure at maturity.
            7. Spikelets glabrous; nodes of culms glabrous; leaves glabrous or variously pubescent, but not velvety on both surfaces_____7. *P. ciliatifolium*
            7. Spikelets pubescent, often densely so; nodes of culms pubescent; leaves velvety on both surfaces_____ _____8. *P. bushii*

1. **Paspalum dissectum** (L.) L. Sp. Pl. ed. 2:81. 1762. *Fig. 256.*
Panicum dissectum L. Sp. Pl. 57. 1753.
Paspalum dimidiatum L. Syst. Nat. ed. 10, 2:855. 1759.
Paspalum walterianum Schult. Mant. 2:166. 1824.
Creeping, branching, glabrous, subaquatic perennial; culms repent, 20–60 cm long, often forming mats; leaves 3–6 cm long, 4–5 mm wide; racemes 1–5, 2–3 cm long, terminal or axillary,

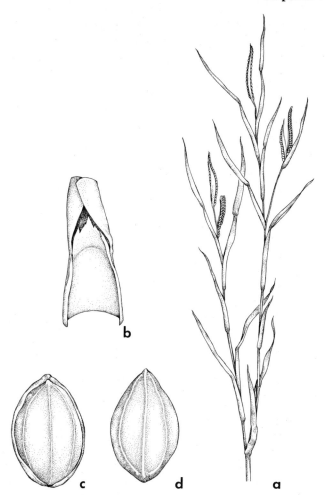

256. *Paspalum dissectum.* *a.* Upper part of plant, X⅓. *b.* Sheath, with ligule, X5. *c.* Spikelet, front view, X15. *d.* Spikelet, back view, X15.

falling entire; rachis membranous, 2–3 mm wide, narrower and shorter than the rows of spikelets but folded over and clasping them; spikelets glabrous, ovoid to obovoid, 2 mm long, 1.4 mm wide; glume and sterile lemma 3- to 5-nerved, slightly longer than the grain; 2n = 40 (Brown, 1951).

HABITAT: Moist soil; edges of shallow swamps.

RANGE: New Jersey to Missouri, south to Texas and Florida.

ILLINOIS DISTRIBUTION: Very rare; known from three counties, and not collected since 1893. The first Illinois collection was made in 1850 by Brendel from St. Clair County.

Paspalum dissectum is distinguished easily by the terminal spikelet extending well beyond the tip of the infolded rachis.

2. Paspalum fluitans (Ell.) Kunth, Rev. Gram. 1:24. 1829. *Fig. 257.*

Ceresia fluitans Ell. Bot. S. C. & Ga. 1:109. 1816.

Paspalum mucronatum Muhl. Descr. Gram. 96. 1817.

Paspalum natans LeConte, Journ. Phys. Chem. 91:285. 1820.

Paspalum frankii Steud. Syn. Pl. Glum. 1:119. 1854.

Sprawling or repent, branching, glabrous, aquatic annual; culms soft and spongy, to 1 m long; leaves 10–20 cm long, 10–15 mm wide; racemes 5–50, usually more than 10, 3–8 cm long, spreading or recurved; rachis herbaceous, 1.3–2.0 mm wide, wider and longer than the rows of spikelets but folded over and clasping them; spikelets minutely glandular-pubescent, ellipsoid, 1.3–2.0 mm long, 0.8 mm wide; glume and sterile lemma 2-nerved, the mid-nerve suppressed, slightly longer than the grain.

COMMON NAME: Swamp Beadgrass.

HABITAT: Floating in shallow standing water. Terrestrial plants are dwarfed.

RANGE: North Carolina to Kansas, south to Texas and Florida.

ILLINOIS DISTRIBUTION: Occasional in the southern two-thirds of the state. The Henderson County record apparently represents the northernmost station for this species in its overall range.

3. Paspalum pubiflorum Rupr. ex Fourn. var. **glabrum** (Vasey) Vasey ex Scribn. Bull. Tenn. Agr. Exp. Sta. 7:32. 1894. *Fig. 258.*

Paspalum remotum var. *glabrum* Vasey, Bull. Torrey Club 13:166. 1886.

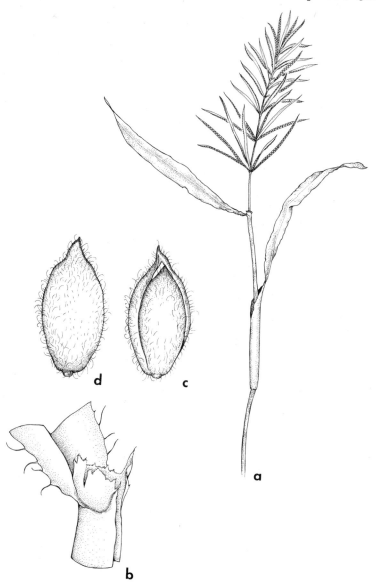

257. *Paspalum fluitans* (Swamp Beadgrass). *a*. Upper part of plant, X½. *b*. Sheath, with ligule, X5. *c*. Spikelet, front view, X17½. *d*. Spikelet, back view, X17½.

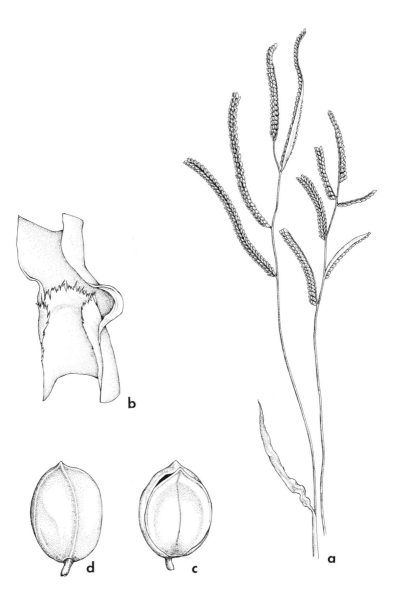

258. *Paspalum pubiflorum* var. *glabrum* (Beadgrass). *a.* Inflorescences, X½. *b.* Sheath, with ligule, X7½. *c.* Spikelet, front view, X10. *d.* Spikelet, back view, X10.

Paspalum geminum Nash. Bull. N. Y. Bot. Gard. 1:434. 1900.
Paspalum laeviglume Scribn. ex Nash in Small, Fl. Southeast.
U. S. 75. 1903.

Decumbent perennial, rooting at the nodes; culms stout, glabrous
to pubescent, geniculate, to 2 m tall; leaves 10–15 cm long, 6–20
mm wide, pilose on the margin; racemes 4–8, usually 5 or more,
the lower frequently distant, 2–10 cm long; rachis with scarious,
nearly winglike margins, 1.2–2.0 mm wide, frequently near maxi-
mum; spikelets paired in double rows, glabrous, obovoid, 3.0–3.2
mm long, about 2 mm wide; glume and sterile lemma 3- to 5-
nerved.

COMMON NAME: Beadgrass.

HABITAT: In moist soil in ditches, along roadsides, and
along streams; tolerant of drought.

RANGE: Pennsylvania to Kansas, south to Texas and
Florida.

ILLINOIS DISTRIBUTION: Occasional in the southern one-
third of the state; absent elsewhere.

Paspalum pubiflorum var. *glabrum* may be distinguished
easily from other Illinois species by the wide rachis and
the paired spikelets which appear to be in four rows
along the rachis.

Nash's *P. geminum*, used by several Illinois workers, is based
upon nearly identical material of *P. pubiflorum* var. *glabrum*.

This grass is sometimes used for forage.

4. **Paspalum floridanum** Michx. Fl. Bor. Amer. 1:44. 1803.
Fig. 259.

Paspalum glabrum Bosc in Flugge, Monogr. Pasp. 172, 1810.
Paspalum laevigatum Poir. Encycl. Suppl. 4:313. 1816.
Paspalum laeve var. *floridanum* (Michx.) Wood, Class-book
782. 1861.
Paspalum floridanum var. *glabratum* Engelm. ex Vasey, Bull.
Torrey Club 13:166. 1886.
Paspalum glabratum (Engelm.) Mohr, Bull. Torrey Club 24:
21. 1897.

Erect, stout, glabrous perennial; culms solitary or few, to 2 m tall;
leaves 12–50 cm long, 4–10 mm wide; racemes 2–6, usually 3 or
4, 4–12 cm long, suberect or ascending; rachis 1.0–1.4 mm wide,
strongly flexuous; spikelets in pairs (one of pair sometimes rudi-

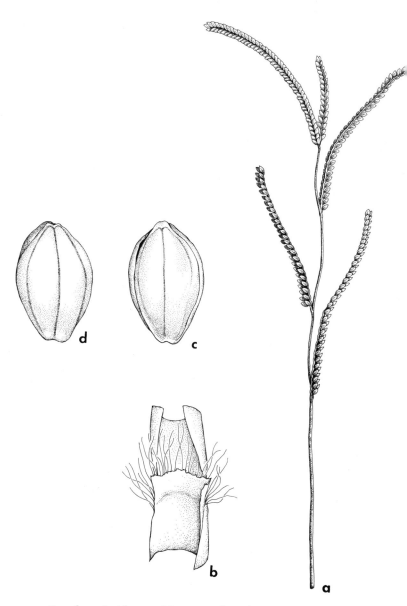

259. *Paspalum floridanum* (Giant Beadgrass). *a.* Inflorescence, X½. *b.* Sheath, with ligule, X5. *c.* Spikelet, front view, X7½. *d.* Spikelet, back view, X7½.

mentary), glabrous, crowded, oval, 3.6–4.0 mm long, 2.8–3.1 mm wide; glume and sterile lemma firm but papery, 5-nerved, scarcely covering grain at maturity; 2n = 160 (Burton, 1942), 120 (Brown, 1948).

COMMON NAME: Giant Beadgrass.

HABITAT: Low, moist sandy soil.

RANGE: Pennsylvania to Kansas, south to Oklahoma, Texas, and Florida.

ILLINOIS DISTRIBUTION: Not common in the southern one-third of the state; absent elsewhere.

The first Illinois collection was made in 1946.

Paspalum floridanum is distinguishable from other Illinois species by its extremely large spikelets.

5. **Paspalum laeve** Michx. Fl. Bor. Amer. 1:44. 1803. *Fig. 260.*

Paspalum tenue Darby, Bot. South. States 576. 1857.

Paspalum circulare Nash in Britton, Man. Fl. N. States 73. 1901.

Paspalum laeve var. *circulare* (Nash) Stone, Ann. Rep. N. J. Mus. 1910; 187. 1911.

Erect or ascending, tufted, glabrous to ciliate or pilose, perennial; culms slender and firm, to 1.3 m tall; leaves 5–30 cm long, 3–10 mm wide; racemes 2–7, usually 3 or 4, 4–17 cm long, ascending or spreading; rachis about 1 mm wide, with a tuft of hairs at base; spikelets solitary, glabrous, suborbicular to orbicular, 2.5–3.2 mm long, 2.0–2.5 mm wide; glume and sterile lemma 5-nerved with the lateral veins approximate at the margins; grain similar in shape and size to spikelet, the tip exposed at maturity; 2n = 40 (Brown, 1948).

HABITAT: Moist soils of roadside ditches, meadows, and stream borders.

RANGE: Massachusetts to Kansas, south to Texas and Florida.

ILLINOIS DISTRIBUTION: Rather common in the southern one-half of the state; absent elsewhere.

Paspalum laeve can be distinguished from other Illinois species by the combination of a narrow rachis and a 5-nerved sterile lemma.

There seems to be little justification in recognizing *P. laeve* and *P. circulare* as distinct taxa since much intergradation

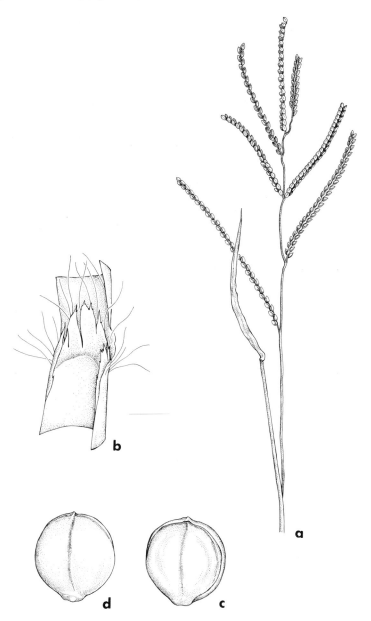

260. *Paspalum laeve.* *a.* Inflorescence, X½. *b.* Sheath, with ligule, X5.
c. Spikelet, front view, X7½. *d.* Spikelet, back view, X7½.

occurs in the size of the spikelets. In their extreme conditions, *P. laeve* has small spikelets which are longer than broad, while *P. circulare* has large spikelets which are orbicular. Both extremes, as well as intermediate forms, occur in Illinois.

6. Paspalum lentiferum Lam. Tabl. Encycl. 1:175. 1791. *Fig. 261.*

Erect, rather robust, usually glabrous culms to 1.5 m tall; leaves to 25 cm long, to 7 mm wide, pilose, the sheaths pilose and strongly keeled; racemes 4–5, spreading-ascending; rachis slender, 1.5–2.0 mm wide; spikelets paired and solitary in the same raceme, suborbicular, 2.7–3.4 mm long, the glume and sterile lemma delicate.

HABITAT: Wet, roadside ditches; low, post oak flats.

RANGE: Virginia to Florida and Texas; Illinois.

ILLINOIS DISTRIBUTION: Rare; known only from Pulaski and Massac counties.

When this southeastern species was first collected from a wet roadside ditch in Pulaski County in 1961, it was considered adventive because of the great gap in its geographic distribution. Later, however, when this species was discovered at the Mermet Conservation Lake both in wet, roadside ditches and in low, post oak flats, a native condition was suspected.

This robust species is recognized by its large suborbicular spikelets, some of which are paired and some of which are solitary in the same raceme.

7. Paspalum ciliatifolium Michx. Fl. Bor. Am. 1:44. 1803. *Fig. 262.*

Paspalum pubescens Muhl. ex Willd. Enum. Pl. 89. 1809.

Paspalum spathaceum Desv. in Poir. in Lam. Encycl. Suppl. 4:314. 1816.

Paspalum ciliatifolium var. *brevifolium* Vasey, Proc. Acad. Phila. 1886:285. 1886.

Paspalum setaceum var. *ciliatifolium* (Michx.) Vasey, Contr. U. S. Nat. Herb. 3:17. 1892.

Paspalum stramineum Nash in Britton, Man. 1:74. 1901.

Paspalum muhlenbergii Nash in Britton, Man. 1:75. 1901.

Paspalum pubescens var. *muhlenbergii* (Nash) House, Bull. N. Y. State Museum 243–44:39. 1923.

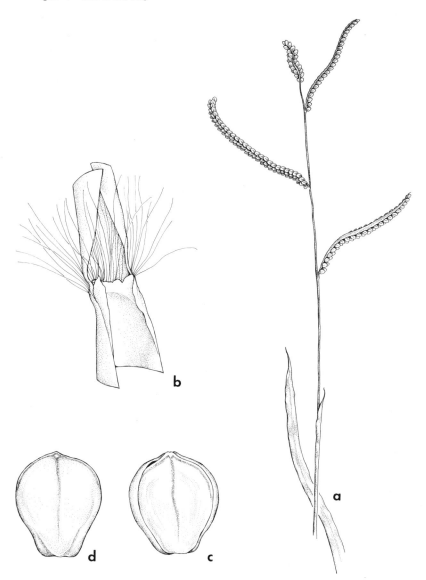

261. *Paspalum lentiferum.* *a.* Inflorescence, X½. *b.* Sheath, with ligule, X5. *c.* Spikelet, front view, X10. *d.* Spikelet, back view, X10.

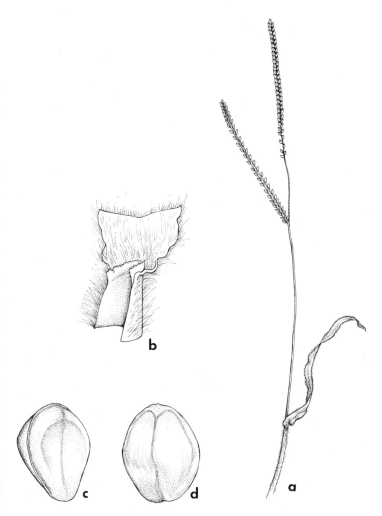

262. *Paspalum ciliatifolium* (Beadgrass). *a.* Inflorescence, X½. *b.* Sheath, with ligule, X5. *c.* Spikelet, front view, X12½. *d.* Spikelet, back view, X12½.

Paspalum ciliatifolium var. *stramineum* (Nash) Fern. Rhodora 36:20. 1934.

Paspalum ciliatifolium var. *muhlenbergii* (Nash) Fern. Rhodora 36:20. 1934.

Erect or spreading, tufted, glabrous to ciliate or pilose, perennial; culms slender, appressed, to 1 m tall, with glabrous nodes; leaves 6–35 cm long, 2–20 mm wide, glabrous or variously pubescent; racemes 1–3 (–4), slender, arching; rachis slender, 1.0–1.2 mm wide, with a tuft of hairs at the base; spikelets in pairs, glabrous, shiny, suborbicular, 1.8–2.4 mm long, 1.3–1.7 mm wide; glume and sterile lemma 3-nerved, the mid-vein frequently obscure; grain similar in size and shape to the spikelet; 2n = 20 (Brown, 1948).

COMMON NAME: Beadgrass.

HABITAT: Dry or moist sandy soils; sometimes in open woods.

RANGE: Vermont to Minnesota, south to Colorado, Arizona, and Florida.

ILLINOIS DISTRIBUTION: Rather common in the southern two-thirds of the state; rare elsewhere.

Paspalum ciliatifolium may be distinguished from all other species of *Paspalum* in Illinois except *P. bushii* by the very small spikelets and the sterile lemma with three nerves. From *P. bushii* it differs by its glabrous spikelets, its glabrous nodes, and its leaves which are not velvety on both surfaces.

Paspalum ciliatifolium is extremely variable in its pubescence, although the types of pubescence seem to intergrade hopelessly. Those specimens in which the leaf surface is glabrous have been designated var. *ciliatifolium;* those in which the upper leaf surface possesses both minute hairs as well as long hairs have been called var. *stramineum* (or *P. stramineum*); those in which the upper leaf surface bears only long hairs of nearly equal length have been known as var. *muhlenbergii* (or *P. muhlenbergii*). None of these varieties is maintained in this work.

8. Paspalum bushii Nash in Britton, Man. 1:74. 1901. *Fig. 263.*
Erect, tufted perennial; culms slender, more or less appressed, to nearly 1 m tall, with pubescent nodes; leaves to 20 cm long, to 15 mm broad, velutinous on both surfaces with both short and long hairs; racemes 2–3, slender, arching; rachis slender, about 1 mm wide, with a tuft of hairs at the base; spikelets in pairs,

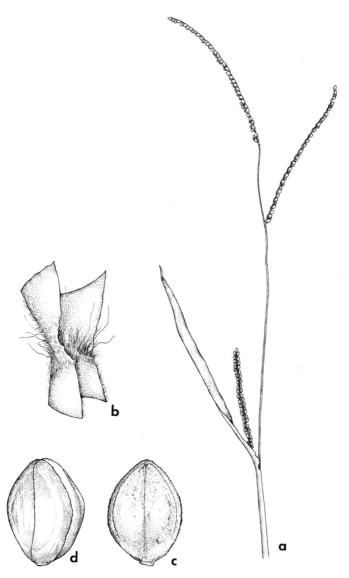

263. Paspalum bushii (Beadgrass). *a.* Inflorescence, X½. *b.* Sheath, with ligule, X5. *c.* Spikelet, front view, X15. *d.* Spikelet, back view, X15.

pubescent, suborbicular, 1.9–2.2 mm long, somewhat narrower; glume and sterile lemma 3-nerved, the nerves frequently obscured by the pubescence; grain similar in size and shape to the spikelet.

COMMON NAME: Beadgrass.

HABITAT: Fields; along edge of woods.

RANGE: Illinois to Nebraska, south to Texas.

ILLINOIS DISTRIBUTION: Restricted to a few southwestern counties.

This species, sometimes considered to be a pubescent variant of *P. ciliatifolium,* is maintained as distinct in this work on the basis of the pubescent spikelets, pubescent nodes, and velvety leaf surfaces. In addition, its very restricted range seems significant.

Species Excluded

Agropyron caninum (L.) Beauv. Although this species is reported by Pepoon (1927) from Cook County, there apparently are no specimens to validate this report.

Agropyron dasystachyum (Hook.) Scribn. Thieret and Evers (1957) point out that all previous reports of this species from Illinois are errors for *A. smithii* var. *molle*. *Agropyron dasystachyum* apparently has not been found in Illinois.

Alopecurus geniculatus L. Although most early Illinois botanists used this binomial for what is now *A. carolinianus*, *Alopecurus geniculatus* refers to a species different from the plants which occur in Illinois.

Ammophila arenaria (L.) Link. This is the binomial given to *A. breviligulata* Fern. by Gates (1912), Mosher (1918), and Pepoon (1927).

Beckmannia erucaeformis (L.) Host. This is the binomial which was given to *B. syzigachne* by Pepoon (1927).

Calamagrostis arenaria (L.) Roth. This is the binomial which was used by many early Illinois workers for *Ammophila breviligulata* Fern. Fernald (1920) points out that *Calamagrostis arenaria* is a different species.

Calamagrostis coarctata Torr. ex Eaton. Although Mead (1846) and Lapham (1857) reported this species from Illinois, their reports are based on misidentifications for *C. canadensis*.

Elymus diversiglumis Scribn. & Ball. This species ranges north of Illinois. The Brendel collection which Mosher (1918) reported as *E. diversiglumis* is *Elymus hystrix* L. with long, setaceous glumes and a more crowded inflorescence.

Festuca nutans Moench. This species was not distinguished from *F. obtusa* by the early workers of Illinois. It appears, however, to be a distinct species which is unknown from the United States.

Holcus halepensis L. This is the binomial which Mosher (1918) used erroneously for *Arrhenatherum elatius* (L.) Presl.

Hordeum pratense Huds. Patterson (1874; 1876) and Schneck (1876) used this binomial erroneously for *H. pusillum*. The true *H. pratense* is a different species.

Paspalum longepedunculatum LeConte. The Mosher (1918)

report of this southeastern species is an error for *P. ciliatifolium* Michx. with minutely hairy leaf surfaces.

Paspalum longipilum Nash. Mosher (1918) reported this species from Illinois, but the specimens on which she based this report are actually *P. pubiflorum* Rupr. ex Fourn. var. *glabrum* (Vasey) Vasey ex Scribn.

Paspalum repens Bergius. The report of this species by Mosher (1918) from Illinois is an error for *P. fluitans* (Ell.) Kunth, since *P. repens* is an entirely different species.

Paspalum supinum (Bosc) Poir. ex Lam. The specimens on which Mosher (1918) based this report are *P. ciliatifolium* Michx. *Paspalum supinum* is a species of the southeastern states.

Poa cuspidata Nutt. in Bart. Since I have seen no Illinois collections of this grass, and since Hitchcock (1950) does not include it from Illinois, I am omitting *Poa cuspidata* from the Illinois flora.

Poa interior Rydb. Although I identified a collection from Piney Creek, Randolph County, as *Poa interior*, and reported this in 1967, I am excluding this species from the Illinois flora since I have not been able to relocate the specimen.

Sphenopholis pallens (Biehler) Scribn. First attributed to Illinois by Deam (1910) and subsequently by other authors, this species occurs southeast of Illinois. The specimens on which the Illinois determinations were made are *S. obtusata* var. *major*.

Sphenopholis palustris (Michx.) Scribn. Reported from Illinois by Robinson and Fernald (1908), this is a synonym for *Trisetum pennsylvanicum* (L.) Beauv. (See latter under Species Excluded.)

Sphenopholis pennsylvanica (L.) Hitchc. Reported by Mosher (1918), this binomial is a synonym for *Trisetum pennsylvanicum* (L.) Beauv. (See latter under Species Excluded.)

Stipa avenacea L. Lapham (1857) describes and illustrates this species, noting that it occurs in dry, sandy places in timbered land and openings. Although this species may well be in Illinois, it is excluded from the flora since no specimens have been seen to substantiate its occurrence in the state.

Trisetum palustre (Michx.) Torr. This is a synonym for *Trisetum pennsylvanicum* (L.) Beauv. (See latter under Species Excluded.)

Trisetum pennsylvanicum (L.) Beauv. The first report of this species from Illinois was in 1856 by Gray as *Trisetum palustre* (Michx.) Torr. Although several later authors likewise reported this species from Illinois, probably copying Gray's reference, there have been no specimens found to authenticate its occurrence in Illinois.

Summary of the Taxa of Grasses in Illinois

	Genera	Species	Lesser Taxa
Festucoideae	(40)	(120)	(18)
Festuceae	8	49	2
Aveneae	18	32	4
Triticeae	7	22	11
Meliceae	3	9	1
Stipeae	2	6	0
Brachyelytreae	1	1	0
Diarrheneae	1	1	0
Panicoideae	(19)	(89)	(23)
Paniceae	9	71	20
Andropogoneae	10	18	3
Eragrostoideae	(20)	(68)	(7)
Eragrosteae	8	39	5
Chlorideae	10	17	0
Aeluropodeae	1	1	0
Aristideae	1	11	2
Bambusoideae	(1)	(1)	(0)
Bambuseae	1	1	0
Oryzoideae	(3)	(5)	(1)
Oryzeae	3	5	1
Arundinoideae	(3)	(3)	(0)
Arundineae	1	1	0
Danthonieae	1	1	0
Centotheceae	1	1	0
Totals	86	286	49

GLOSSARY
LITERATURE CITED
INDEX OF PLANT NAMES

GLOSSARY

Acuminate. Gradually tapering to an elongated point.

Acute. Sharp, ending in a point.

Annual. Living for a single year.

Anther. The terminal part of a stamen which bears the pollen.

Antrorse. Pointing upward.

Apiculate. Ending abruptly in a small, sharp tip.

Appressed. Lying flat against the surface.

Aristate. Bearing an awn.

Attenuate. Long-tapering.

Auriculate. Bearing an ear-like process.

Awn. A bristle usually terminating a structure.

Axis. The central support to which lateral parts are attached.

Bidentate. Having two teeth.

Bifid. Two-cleft.

Callus. A hard swollen area at the outside base of a lemma or palea.

Canescent. Grayish-hairy.

Capillary. Threadlike.

Carinate. Bearing a keel.

Cartilaginous. Firm but flexible.

Caryopsis. A type of one-seeded, dry, indehiscent fruit with seed coat attached to the mature ovary wall.

Caudex. (pl., **caudices**). The woody base of a perennial plant.

Cauline. Belonging to a stem.

Cespitose. Growing in a tuft.

Chartaceous. Papery.

Ciliate. Bearing marginal hairs.

Compressed. Flattened.

Conduplicate. Folded together lengthwise.

Connate. United, when referring to like parts.

Connivent. Coming in contact; converging.

Convex. Rounded on the outer surface; opposite of concave.

Coriaceous. Leathery.

Culm. The stem which terminates in an inflorescence.

Cuspidate. Terminating in a very short point.

Decumbent. Lying flat, but with the top ascending.

Diffuse. Loosely spreading.

Digitate. Radiating from a common point, like the fingers from a hand.

Dioecious. With staminate flowers on one plant, pistillate flowers on another.

Disarticulate. To come apart; to become disjointed.

Divergent. Spreading apart.

Ellipsoid. Referring to a solid object which, in side view, is broadest at the middle, gradually tapering equally to both ends.

Elliptic. Broadest at the middle, gradually tapering equally to both ends.

Emarginate. Deeply notched at the tip.

Erose. With an irregularly notched margin.

Fascicle. A cluster; a bundle.

Fibrous. Bearing fibers; i.e., slender projections of equal diameters.

Filiform. Threadlike.

Flexuous. Zigzag.

Floret. A small flower.

Geniculate. Bent.

Glabrate. Becoming smooth.

Glabrous. Smooth; without hairs, scales, or glands.

Glaucous. With a whitish covering which can be rubbed off.

Glume. A sterile scale subtending a spikelet.

Grain. The fruit of most grasses.

Hirsute. With stiff hairs.

Hirtellous. With minute stiff hairs.

Hispid. With rigid hairs.

Hispidulous. With minute rigid hairs.

Hyaline. Transparent.

Indurate. Hardened.

Inflorescence. A cluster of flowers.

Internode. The area between two consecutive nodes.

Involute. Rolled inward.

Keel. A central ridge.

Lanceolate. Lance-shaped; broadest near base, gradually ta-

pering to the narrow apex.

Lanceoloid. Referring to a solid object which is broadest near base, gradually tapering to the narrow apex.

Lemma. A scale subtending the floret.

Ligule. The structure on the inner surface of the leaf at the junction of the blade and the sheath.

Linear. Elongated and uniform in width throughout.

Lodicule. A small rudimentary structure at the base of a grass flower.

Monoecious. With stamens and pistils in separate flowers on the same plant.

Mucronate. Bearing a short, terminal point.

Nerve. Vein.

Node. That place on the stem from which leaves and branchlets arise.

Oblong. With nearly uniform width throughout, but broader than linear.

Oblongoid. Referring to a solid object which, in side view, is nearly uniform in width throughout.

Obovate. Broadly rounded at apex, becoming narrowed below; broader than oblanceolate.

Obsolete. Not apparent.

Obtuse. Rounded; blunt.

Orbicular. Round.

Ovary. The lower swollen part of the pistil which produces the ovules.

Ovoid. Referring to a solid

object which, in side view, is broadly rounded at base, becoming narrowed above.

Ovule. The egg-producing structure found within the ovary; an immature seed.

Palea. The scale opposite the lemma which encloses the flower.

Panicle. A type of inflorescence composed of several racemes.

Papillose. Bearing pimplelike processes.

Pedicel. The individual stalk of a spikelet.

Pedicellate. Bearing a pedicel.

Peduncle. The stalk of an inflorescence.

Perennial. Living more than one year.

Perfect. Bearing both stamens and pistils.

Perianth. That part of the flower composed of the calyx or corolla or both.

Pericarp. The ripened ovary wall.

Pilose. Bearing soft long hairs.

Pistil. Female reproductive organ.

Plicate. Folded.

Prostrate. Lying flat.

Puberulent. Minutely pubescent.

Raceme. A type of inflorescence where pedicellate flowers are arranged along an elongated axis.

Racemose. Bearing racemes.

Rachilla. The axis bearing the flowers.

Rank. Referring to the number of planes in which structures are borne.

Reflexed. Turned downward.

Retrorse. Pointing downward.

Retuse. Shallowy notched at a rounded apex.

Rhizomatous. Bearing rhizomes.

Rugose. Wrinkled.

Rugulose. With small wrinkles.

Scaberulous. Slightly rough to the touch.

Scabrous. Rough to the touch.

Scarious. Thin and membranous.

Sericeous. Silky; bearing soft, appressed hairs.

Serrate. With teeth which project forward.

Serrulate. With very small teeth which project forward.

Sessile. Without a stalk.

Seta. Bristle.

Setose. Bearing setae.

Setulose. Bearing small setae.

Sheath. A protective covering; the basal part of a grass leaf that encircles the stem.

Spicate. Bearing a spike.

Spike. A type of inflorescence where sessile flowers are arranged along an elongated axis.

Spikelet. The basic unit in a grass inflorescence.

Spinulose. With small spines.

Stamen. The male reproductive organ.

Staminate. Bearing stamens.

Stigma. The apex of the pistil which receives the pollen.

Stipitate. Bearing a stipe or stalk.

Stolon. A slender, horizontal stem on the surface of the ground.

Stoloniferous. Bearing stolons.

Strigose. With appressed, straight hairs.

Style. That elongated part of the pistil between the ovary and the stigma.

Subulate. With a very short, narrow point.

Terete. Round in cross section.

Translucent. Partly transparent.

Truncate. Abruptly cut across.

Umbonate. With a stout projection at the center.

Villous. With long, soft, slender, unmatted hairs.

Viscid. Sticky.

Whorled. An arrangement of three or more structures at a point on the stem.

LITERATURE CITED

Aase, H. C. and L. R. Powers. 1926. Chromosome numbers in crop plants. American Journal of Botany 13:367–72.

Anderson, D. E. 1961. Taxonomy and distribution of the genus *Phalaris*. Iowa State College Journal of Science 36:1–96.

Armstrong, J. M. 1937. A cytological study of the genus *Poa*. Canadian Journal of Research, C. 15:281–97.

Avdulov, N. P. 1928. Karyo-systematische Untersuchungen der Familie Gramineen. All Union Cong. Bot. Moscow Jour. 65–67.

———. 1931. Karyo-systematische Untersuchungen der Familie Gramineen. Bulletin of Applied Botany, Genetics and Plant Breeding, Supplement 44. 428 pp.

Bowden, W. M. 1957. Cytotaxonomy of the section Psammelymus of the genus *Elymus*. Canadian Journal of Botany 35:951–92.

Boyle, W. S. and A. H. Holmgren. 1955. A cytogenetic study of natural and controlled hybrids between *Agropyron trachycaulum* and *Hordeum jubatum*. Genetics 40:539–45.

Brown, W. L. 1939. Chromosome complements of five species of *Poa* with an analysis of variation in *Poa pratensis*. American Journal of Botany 26:717–23.

Brown, W. V. 1948. A cytological study in the Gramineae. American Journal of Botany 35:382–95.

———. 1950. A cytological study of some Texas Gramineae. Bulletin of the Torrey Botanical Club 77:63–76.

———. 1951. Chromosome numbers of some Texas grasses. Bulletin of the Torrey Botanical Club 78:292–99.

———. 1958. Leaf anatomy in grass systematics. Botanical Gazette 119:170–78.

Church, G. L. 1929. Meiotic phenomena in certain Gramineae. I. Festuceae, Aveneae, Agrostideae, Chlorideae, and Phalarideae. Botanical Gazette 87:608–29.

———. 1936. Cytological studies in the Gramineae. American Journal of Botany 23:12–15.

———. 1949. A cytotaxonomic study of *Glyceria* and *Puccinellia*. American Journal of Botany 36:155–65.

———. 1954. Interspecific hybridization in eastern *Elymus*. Rhodora 56:185–97.

———. 1967. Taxonomic and genetic relationships of eastern North American species of *Elymus* with setaceous glumes. Rhodora 69:121–62.

Church, G. L. 1967a. Pine Hills *Elymus*. Rhodora 69:330–51.

Clausen, R. T. 1952. Suggestion for the assignment of *Torreyochloa* to *Puccinellia*. Rhodora 54:42–45.

Covas, G. 1949. Taxonomic observations on the North American species of *Hordeum*. Madroño 10:1–21.

Cugnac, A. de and M. Simonet. 1941. Les nombres de chromosomes de quelques espècies du genre *Bromus* (Gramineae). Comp. Rend. Sociète Biologie Paris 135:728–31.

Elliott, F. C. 1949. Cross-fertility and cytogenetics of selected Bromopsis section members within the genus *Bromus* L. Iowa State College Journal of Science 24:44–45.

Erdman, K. S. 1965. Taxonomy of the genus *Sphenopholis* (Gramineae). Iowa State Journal of Science 39:289–336.

Fernald, M. L. 1920. The American *Ammophila*. Rhodora 22:70–71.

———. 1928. The American and eastern Asiatic *Beckmannia*. Rhodora 30:27–34.

———. 1930. The identity of *Alopecurus aequalis*. Rhodora 32:221–22.

———. 1933. Types of some American species of *Elymus*. Rhodora 35:187–98.

———. 1935. The allies of *Festuca ovina* in eastern America. Rhodora 37:250–52.

———. 1945. Botanical specialities of the Seward Forest and adjacent areas of southeastern Virginia. Rhodora 47:107–8.

———. 1945b. An incomplete flora of Illinois. Rhodora 47:204–19.

———. 1950. Gray's Manual of Botany. 8th ed. New York: American Book Company. 1632 pp.

Gates, F. C. 1912. The vegetation of the beach area in northeastern Illinois and southeastern Wisconsin. Bulletin of the Illinois State Laboratory of Natural History 9:255–372.

Glassman, S. F. 1964. Grass flora of the Chicago region. The American Midland Naturalist 72:1–49.

Gleason, H. A. 1910. The vegetation of the inland sand deposits of Illinois. Bulletin of the Illinois State Laboratory of Natural History 9:23–174.

———. 1952. The New Britton and Brown Illustrated Flora of the Northeastern United States and Adjacent Canada. I. New York: The New York Botanical Garden. 590 pp.

Gould, F. 1968. Grass Systematics. New York: McGraw-Hill Book Company. 382 pp.

Gray, A. 1856. A Manual of the Botany of the Northern United States. 2nd ed. New York.

Gross, A. T. H. 1960. Distribution and cytology of *Elymus macounii* Vasey. Canadian Journal of Botany 38:63–67.

Hansen, A. A. and H. D. Hill. 1953. The meiotic behavior of hexaploid orchard grass (*Dactylis glomerata* L.). Bulletin of the Torrey Botanical Club 80:113–22.

Hartung, M. E. 1946. Chromosome numbers in *Poa, Agropyron,* and *Elymus.* American Journal of Botany 33:516–31.

Henrard, J. T. 1950. Monograph of the genus *Digitaria.* Leiden: Universitare Pers Leiden.

Hitchcock, A. S. 1920. The Genera of Grasses of the United States, with special reference to the economic species. United States Department of Agriculture Bulletin 772.

———. 1925. The North American species of *Stipa.* Contributions to the United States National Herbarium 24:215–62.

———. 1935. Manual of the Grasses of the United States. United States Department of Agriculture Miscellaneous Publication Number 200.

———. 1950. Manual of the Grasses of the United States. 2nd ed. revised by Agnes Chase. United States Department of Agriculture Miscellaneous Publication. 1051 pp.

Hutchinson, J. 1959. The Families of Flowering Plants. II. Monocotyledons. Oxford: Clarendon Press. 792 pp.

Jenkin, T. J. and P. T. Thomas. 1938. The breeding affinities and cytology of *Lolium* species. Journal of Botany 76:10–12.

Johnson, B. L. 1945. Cytotaxonomic studies in *Oryzopsis.* Botanical Gazette 107:1–32.

——— and G. A. Rogler. 1943. A cytotaxonomic study of an intergeneric hybrid between *Oryzopsis hymenoides* and *Stipa viridula.* American Journal of Botany 30:49–56.

Johnsson, H. 1941. Cytological studies in the genus *Alopecurus.* Acta Universit. Lund 37:no. 3.

Jones, G. N. 1945. Flora of Illinois. 1st ed. South Bend, Indiana: University of Notre Dame Press. 317 pp.

———. 1950. Flora of Illinois. 2nd ed. South Bend, Indiana: University of Notre Dame Press. 368 pp.

——— et al. 1955. Vascular Plants of Illinois. Urbana: The University of Illinois Press, and the Illinois State Museum, Springfield. 593 pp.

Kattermann, G. 1933. Weitere zytologische Untersuchungen an *Briza media* mit besonderer Beruck sichtigung der durch Verbande aus vier Chromosomen ausgezeichneten Pflanzen. Jahrbucher Wiss. Bot. 73:43–91.

Knowles, P. F. 1944. Interspecific hybridizations of *Bromus.* Genetics 29:128–40.

Lapham, I. A. 1857. The native, naturalized, and cultivated grasses of the state of Illinois. Transactions of the Illinois State Agricultural Society 2:551–613.

Lawrence, W. E. 1945. Some ecotypic relations of *Deschampsia caespitosa.* American Journal of Botany 32:298–314.

Marchal, E. 1920. Recherches sur les variations numèriques des chromosomes dans la série végétale. Mem. Acad. Royal Belg. 2, 4:1–108.

Mead, S. B. 1846. Catalogue of plants growing spontaneously in the state of Illinois, the principal part near Augusta, Hancock County. Prairie Farmer 6:35–36, 60, 93, 119–22.

Mohlenbrock, R. H. and J. E. Ozment. 1967. Additions to the grass flora of Illinois. Transactions of the Illinois State Academy of Science 60:184–85.

Mohlenbrock, R. H. 1970. The Illustrated Flora of Illinois. Flowering Plants: Flowering Rush to Rushes. Carbondale: Southern Illinois University Press. 272 pp.

Mosher, E. 1918. The grasses of Illinois. Bulletin of the Agricultural Experiment Station, University of Illinois 205:257–425.

Müntzing, A. 1937. Polyploidy from twin seedlings. Cytologia, Fuj. Jub. Vol. 211–27.

Myers, W. M. 1944. Cytological and genetical analysis of chromosomal association and behavior during meiosis in hexaploid timothy (*Phleum pratense* L.). Journal of Agricultural Research 68:21–33.

———. 1947. Cytology and genetics of forage grasses. Botanical Review 13:319–421.

——— and H. D. Hill. 1947. Distribution and nature of polypoidy in *Festuca elatior* L. Bulletin of the Torrey Botanical Club 74:99–111.

Nielsen, E. L. 1939. Grass Studies III. Additional somatic chromosome complements. American Journal of Botany 26:366–72.

Nygren, A. 1954. Investigations on North American *Calamagrostis*. Hereditas 40:375–97.

Oestergren, C. 1942. Chromosome numbers in *Anthoxanthum*. Hereditas 28:242–43.

Patterson, H. N. 1874. A list of plants collected in the vicinity of Oquawka, Henderson County. Oquawka, Illinois. 18 pp.

———. 1876. Catalogue of the phaenogamous and vascular cryptogamous plants of Illinois. Oquawka, Illinois. 54 pp.

Pepoon, H. S. 1927. An annotated flora of the Chicago area. Chicago: The Chicago Academy of Sciences. 554 pp.

Peto, F. H. 1933. The cytology of certain intergeneric hybrids between *Festuca* and *Lolium*. Journal of Genetics 28:113–56.

Philp, J. 1933. The genetics and cytology of some interspecific hybrids of *Avena*. Journal of Genetics 27:133–79.

Piper, C. V. 1906. North American species of *Festuca*. Contributions to the United States National Herbarium 10:1–48.

Reeder, J. 1957. The embryo in grass systematics. American Journal of Botany 44:756–68.

Robinson, B. L. and M. L. Fernald. 1908. Gray's New Manual of Botany. 7th ed. New York: American Book Company.

St. John, H. 1915. *Elymus arenarius* and its American representatives. Rhodora 17:98–103.

St. Yves, A. 1926. Contribution à l'étude des *Festuca* (subgenus Eufestuca) de l'Amerique du nord et du Mexico. Candollea 2:229–316.

Schneck, J. 1876. Catalogue of the flora of the Wabash Valley. Annual Report of the Geological Survey of Indiana 7:504–79.

Scribner, F. L. 1906. The genus *Sphenopholis*. Rhodora 8:137–46.

Shear, C. L. 1900. A revision of the North American species of *Bromus* occurring north of Mexico. Bulletin of the United States Department of Agriculture, Division of Agrostology 23:1–66.

Sokolovskaja, A. P. 1938. A caryo-geographical study of the genus *Agrostis*. Cytologia 8:452–67.

Stahlin, A. 1929. Morphologische und zytologische Untersuchungen an Gramineen. Pflanzenbau 1:330–98.

Stebbins, G. L. 1930. A revision of some North American species of *Calamagrostis*. Rhodora 32:35–57.

———— and B. Crampton. 1961. A suggested revision of the grass genera of temperate North America. Recent Advances in Botany 1:133–45.

———— and R. M. Love. 1941. A cytological study of California forage grasses. American Journal of Botany 28:371–82.

———— and H. A. Tobgy. 1944. The cytogenetics of hybrids in *Bromus*. American Journal of Botany 31:1–11.

Tanzi, S. 1925. Chromosome numbers of wild barley. Botanical Magazine of Tokyo 39:55–57.

Terrell, E. E. 1967. Meadow fescue: *F. elatior* L. or *F. pratensis* Hudson? Brittonia 19:129–32.

Thieret, J. W. and R. A. Evers. 1957. Notes on Illinois grasses. Rhodora 59:123–24.

Thomas, P. T. 1936. Genotypic control of chromosome size. Nature 138:402.

Thorne, R. F. 1968. Synopsis of a putatively phylogenetic classification of the flowering plants. Aliso 6:57–66.

Vasey, G. 1861. Additions to the flora of Illinois. Transactions of the Illinois Natural History Society 1:139–43.

Verts, B. J. and R. H. Mohlenbrock. 1966. The Illinois taxa of *Paspalum*. Transactions of the Illinois State Academy of Science 59:29–38.

Voss, E. G. 1966. Nomenclatural notes on monocots. Rhodora 68:435–63.

Wagnon, H. K. 1952. A revision of the genus *Bromus*, section Bromopsis, of North America. Brittonia 7:415–80.

Wiegand, K. M. 1918. Some species and varieties of *Elymus* in eastern North America. Rhodora 20:81–90.

Wilson, F. D. 1963. Revision of *Sitanion* (Triticeae, Gramineae). Brittonia 15:303–23.

Winterringer, G. S. and R. A. Evers. 1960. New Records for Illinois Vascular Plants. Scientific Papers Series. II. The Illinois State Museum, Springfield.

Wulff, H. D. 1937. Chromosomenstudien an der schleswig-holsteinischen Angiospermen—Flora I. Berichte Deutsche Botanische Gesellschaft 55:262–69.

INDEX